D0765464

MESOMETEOROLOGY   AT   EL PASO

SCIENCE SERIES   Number Three

o0o

Cover photograph by Charles H. Binion

MESOMETEOROLOGY AT EL PASO

By WILLIS L. WEBB

TEXAS WESTERN PRESS
THE UNIVERSITY OF TEXAS AT EL PASO

1971

CONTENTS

PREFACE

Mesometeorology refers to those dynamic aspects of atmospheric processes which have intermediate scales, smaller
than the *synoptic* phenomena of common weather maps and larger than those parameters sampled at a local weather station.
Included are small-scale phenomena, such as drainage winds
and lee waves induced by local terrain effects, and convective systems, such as thunderstorms and tornadoes. These
dimensions are of the same order as those of a metropolitan
area, so mesometeorology has its principal future application in these populated regions.
    Air pollution problems have focused attention on the
mesometeorological characteristics of our modern cities.
From the crude first look in effect today for elimination
of gross sources of atmospheric pollution, it is clear that
application of mesometeorology must expand to provide input
for systematic control of all industrial effluents (gases,
particles, heat), control of high-speed freeway traffic and
aircraft, and safety applications associated with accidents
involving toxic materials, fires, epidemics and other dis-
asters. A modern city requires advanced planning and sophisticated application techniques to assure that atmospheric
effects are efficiently considered in regulating development and applying safety measures. In addition, future
efforts at weather modification will raise innumerable
problems in applying controls to this most important aspect
of man's environmental resources.
    Meteorology has played an important role in the scien-
tific development of the El Paso area. The presence of
White Sands Missile Range has fostered a rapid expansion of
this scientific field until today the atmosphere of the El
Paso area is the best observed in the world. A notable short-
coming in the meteorological program of the orographically
complex El Paso area is the lack of an adequate boundary-
layer observational and forecast system. The need for and
an outline of such a system are presented here.
    Until the Elephant Butte Reservoir was completed in
1915 *floods* were matters of general concern for residents
of the Rio Grande Valley. Even today local flooding may
result from heavy rains in the surrounding mountain areas.
Floods, along with their alternate severe weather phenomenon,
dust storms, were the focus of meteorological attention in
the early days of meteorology in El Paso.
    Both of these extreme weather phenomena are very infre-
quent, with *sunshine* the principal weather characteristic
of the El Paso area. This good weather has resulted in
strong shifts in emphasis in local meteorology, first from
an agrarian orientation to aviation, and more recently to
rocketry. These technological developments have produced

a wealth of meteorological experience in the El Paso area,
with the locally initiated Meteorological Rocket Network
focusing international attention on progress in the atmospheric sciences in the local area.

In the process of expanding our meteorological fron-
tiers upward in the global sense, small-scale phenomena of
significant impact on the ecology of the El Paso area have
been somewhat neglected. Social pressure relative to air
pollution and weather modification has recently reoriented
our attention to the surface layers to assure that the
technological system which we enjoy does not *excessively
impair* the human characteristics of our environment.

The following pages summarize past events of meteorological significance in the El Paso area, illustrate the
state of knowledge of local atmospheric structure, and de-
pict the structure of a local meteorological observational
system which will be adequate for control of future metropolitan growth. This latter effort at experimental design
is necessarily based on estimates of future technological
demands, and there have been no intentional underestimates
incorporated into the plan. It is clearly time to take
care of the atmospheric aspects of our minimum ecological
needs in an energetic fashion. The plan presented here is
not necessarily the best which can be devised. It does
have the virtue of offering a solid framework on which improvements can be implemented.

The *thesis* that man's control of his environment can be
resolved by legislation for or against specific sources of
atmospheric modification is a gross simplification. Man's
desire for material things and his rapidly increasing capa-
bilities will force our industrial technology to press ever
closer to the upper limits of acceptable local and global
environmental pollution with contaminating gases, particulates, radiations and sounds.

Intelligent control and use of this modification of the
atmosphere can come only from an adequate knowledge of the
impact of pollutants on the El Paso ecology system and an
adequate knowledge of the atmospheric processes which control the contamination cycle. We are getting the former
knowledge through trial and error. We should get the latter
in a *rigorous* scientific manner. An adequate mesoscale
meteorological network in the El Paso area is a technological
necessity.

This treatise includes the material contained in the
masters thesis titled "Mesometeorology of Air Pollution at
El Paso" (1970) at the Meteorology Department, University
of Oklahoma. The author is indebted to Prof. E. H. Antone,
Editor of the Texas Western Press, for his efforts in publication of this book; Prof. Robert L. Schumaker and Philip
H. Duran of the Computational Center at the University of
Texas at El Paso for adapting and operating the computer
programs and to the Physics Department for supporting this
work. Thanks are especially due Mrs. Annette Fishburn for

editing and Mrs. Glenda McMath for her excellent preparation
of the final copy.

1 June 1971   Willis L. Webb

                               Physics Department
                               University of Texas at El Paso
                               El Paso, Texas 79999

                               Atmospheric Sciences Laboratory
                               US Army Electronics Command
                               White Sands Missile Range
                               New Mexico 88002

Chapter I

EL PASO METEOROLOGY

El Paso (del Norte) is located on the Texas-Mexico
border where the Rio Grande alters its southward course
from the Southern Colorado headwaters and flows southeast-
ward across the foothills of the Rockies toward the Gulf of
Mexico as is illustrated in Figure 1.1. For an excellent
historical development of the region the reader is referred
to C. L. Sonnichsen's (1968) book *Pass of the North*. The
Franklin Mountains extend southward to the edge of downtown
El Paso, physically separating the city into a northwest
upper valley section and a northeast branch which stretches
toward the Tularosa Bolson. The Franklins provide an excellent transmitter location (*the highest in Texas*), but,
from a meteorological point of view, they introduce gross
complications into the meteorological processes which es-
tablish the atmospheric structure of The Pass region.

The water which flows through the Rio Grande Channel
is of primary importance to the ecology of this high-altitude
desert region. Strong fluctuations in the streamflow were
principal characteristics of the early uncontrolled days.
Construction of the Elephant Butte Dam approximately one
hundred miles north has removed the fun of river forecast-
ing, reduced the general flooding of lowlands to negligible
proportions and cast most of the streamflow into concrete
irrigation ditches. The impact of this major change in
physiography on the ecology of the area has been complicated
by intensive irrigation efforts, and the terminal ecological
situation which current technological changes will impose
is yet to be determined.

Meteorology in the El Paso area shifted focus away from
the early agricultural emphasis with arrival of the *aviation
age* in the 1930's. Atmospheric observational programs expanded from the surface upward through the troposphere by
means of balloon-borne sensors which attained mean burst
altitudes of 30 km (∿100,000 feet) in the 1960's. In the
mid-1940's, the *rocket age* arrived at El Paso with initiation of rocket testing at White Sands Missile Range. The
tools for exploration of the upper atmosphere came hand in
hand with the requirements for more knowledge of that region.
WAC Corporals, V-2s, Vikings and Aerobees (Newell, 1959)
quickly made White Sands Missile Range the world center for
upper atmospheric exploration. Start of the International
Geophysical Year period in 1958 expanded rocket exploration
to Fort Churchill, Canada, and other locations and precipitated a general interest in synoptic exploration of the
upper atmosphere.

This interest, along with technological advances, primed
the El Paso area in the late 1950's for initiation of the
global synoptic Meteorological Rocket Network (MRN) which

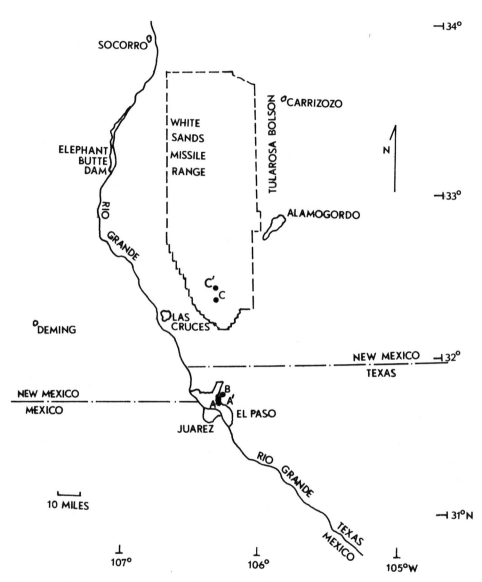

Figure 1.1. Map of the Paso del Norte area.

effectively doubled the observed volume of the atmosphere
(Webb, 1966a). Balloon peak altitudes for meteorological
observations were quickly increased to roughly 60 km
(∿200,000 feet) by introduction of small rocket systems.
The MRN has grown during the first decade of its existance
to include some 29 stations scattered around the world
(Webb, 1969a), with White Sands Missile Range the center for
development of this *global synoptic network* (Webb et al.,
1961, 1962, 1966).

The Atmospheric Sciences Laboratory of White Sands Missile Range has worked with the Schellenger Research Laboratories of the University of Texas at El Paso to play a leading role in development of the required small rocket measurement techniques, data analysis and publication formats which have facilitated this growth of the MRN. Small rocket systems of the Arcas and Loki types have formed the backbone of this new synoptic network, and the wind and temperature sensors (Ballard, 1967) which have been developed for use with these rocket systems have provided revolutionary new concepts of the structure of the upper atmosphere. A total of 3903 stratospheric meteorological soundings had been completed at White Sands Missile Range by the end of 1970, out of a total of more than 13,000 soundings collected from all over the world by the MRN. By the end of 1968 a total of 89 volumes containing more than 11,000 soundings had been assembled, reduced in a common format and published in digital and graphical formats by the Schellenger Research Laboratories. All of these data are also available on punched cards and magnetic tapes for automatic data processing.

Graduate training of scientists for the Meteorological Rocket Network effort was accomplished by the Physics Department of the University of Texas at El Paso. Evening classes were conducted for more than 300 local atmospheric science students, and Summer Institutes were conducted (during the summers of 1966, 1967, 1968, and 1970) for representatives of the several groups involved in the North American Meteorological Rocket Network. This latter effort was expanded by the United States National Academy of Sciences through the Committee on Space Research of the International Council of Scientific Unions to include seminars at the Imperial College of Science and Technology, London, in 1967 and at the Japan Meteorological Agency, Tokyo, in 1968 to get the Global Meteorological Rocket Network under way (Webb, 1969). In addition, the University of Texas at El Paso hosted five national meetings of the *American Meteorological Society* (1958, 1961, 1963, 1966 and 1968) which were designed to consider techniques, air results and foster development of synoptic exploration of the upper atmosphere.

These efforts at exploration of the upper atmosphere through synoptic techniques have made the El Paso area the center of meteorological exploration of the upper atmosphere. This approach is progressing at an accelerated rate with development of new measuring techniques such as the meteor trail radar system (Webb, 1971) which will push the upper limit of synoptic wind observations upward another 30 km to the 110 km level. These accomplishments have made the El Paso area the best observed atmospheric region in the world from a meteorological viewpoint.

While the pollution of the atmosphere has important *global aspects*, the growing importance of urban air pollution precipitated by our advancing technology and growing population has brought to the forefront yet another aspect of the complex problems of the atmospheric sciences. There

are now developing far more stringent requirements for mesoscale meteorological data related to the urban air pollution
control problem, and it is generally true that meteorological
data which had been satisfactory for agricultural and transportation problems are totally inadequate for resolution
of the complex physical processes which determine air
quality.

Limited rainfall and temperature data are available for
several locations in the metropolitan El Paso area, but
these records generally lack the continuity and comparable
exposure which is so necessary for systematic study of the
atmosphere. As an example, rainfall records have been maintained at several Fire Department station locations, but
generally only for those circumstances in which the rainfall
amounts to more than one-quarter inch. While these records
are adequate for their basic purpose of evaluating flooding,
they are hardly adequate for a climatological analysis of
the impact of terrain or air pollution on the precipitation
process.

Temperature and wind data in the El Paso area are even
less satisfactory, primarily as a result of problems in
exposure. While such records do exist, a reasonable evaluation is that they provide for only a crude estimate of
the synoptic variability and climatological structure of the
atmosphere in the El Paso area.

It is then essential that a mesometeorological observational system be implemented in order that the necessary
data for projection of city development be available when the
requirements for that data become critical. The stopgap
measures now in vogue to correct excessive modification of
the atmospheric environment through massive engineering re-
straints on our industrial system will necessarily yield
to more sophisticated application of atmospheric data and
processes for a truly equitable solution to our air pollution problems. A proposal is outlined here to integrate
this need for atmospheric data with the local educational
system, facilitating expansion of student training in the
atmospheric sciences while accumulating the necessary data.

While this approach is but one among many, it has the
virtue of being workable and at the same time bringing the
significance of this field of science to the attention of a
considerable fraction of the future leaders of industry.
In addition, this plan is designed to afford an easy inter-
face with the growing global meteorological system based on
satellite observation and automated data processing and
display. In any case, the plan outlined here offers a base
for concrete alternate proposals to be formulated.

Possibly more important, this plan is designed to facilitate the development of a *total ecological treatment* of
the human urban environment along the lines indicated by
Eddy (1969) through providing an adequate base of knowledge
of the local meteorological structure. Such action is
clearly required before other aspects of the ecological problem can be handled accurately.

GEOGRAPHIC SITUATION

The El Paso metropolitan area is situated along the
northeastern bank of the Rio Grande River in the *Paso del
Norte* region. The city extends northward east of the Franklin Mountains toward the Tularosa Bolson. Juarez, Mexico
is located immediately across the river from El Paso, form-
ing a contiguous metropolitan area of the order of a million
population which is separated only by the political boundary
between the United States and Mexico. Interchange across
this border is growing rapidly, so that the significance
of political boundaries would appear to be easing toward
the very free viewpoint which meteorological processes take
of the area.

As is illustrated in the map of Figure 1.1, White Sands
Missile Range is located a short distance north of El Paso
in the Tularosa Basin. The proximity of this advanced
missile test facility has resulted in assembly of a very
diverse group of atmospheric scientists into the El Paso
area and has placed this locale in the very unique position
of having energetic explorers of all facets of the atmospheric sciences actively working in the immediate area.

The physical structure of the earth's surface in the
El Paso area is illustrated by the general relief map of
Figure 2.1. For additional details, the reader is referred
to (Strain, 1966; Sayre and Livingston, 1945). The river
bed falls smoothly approximately 100 meters (∼300 feet) as
it passes through the region indicated, and both above and
below El Paso flatlands between one-half and two kilometers
in width provide fertile irrigated farm lands. The river
bed is constricted to its narrowest local dimensions in the
downtown El Paso area, with slopes climbing to above 1000
meter heights within a few kilometers on the northeast and
southwest sides of the river. The mountain range provides
a wide range of elevations for the local area and forms a
barrier for the zonal circulation, forcing winds to ease
through the pass when pressure gradients are low and to
execute strong wave and turbulent motions when the winds
become strong enough to require direct passage over the
mountain barrier.

The first case is essentially the *air pollution case*,
since it is under low surface pressure gradient situations
that pollutants can accumulate sufficiently to constitute
a problem. This happens during the fall and early winter
months, and air pollution conditions in El Paso usually
involve a slow drift of surface air through the pass rather
than totally stagnant conditions. This weak surface flow
is usually from the west, although under some conditions
this average condition may be reversed.

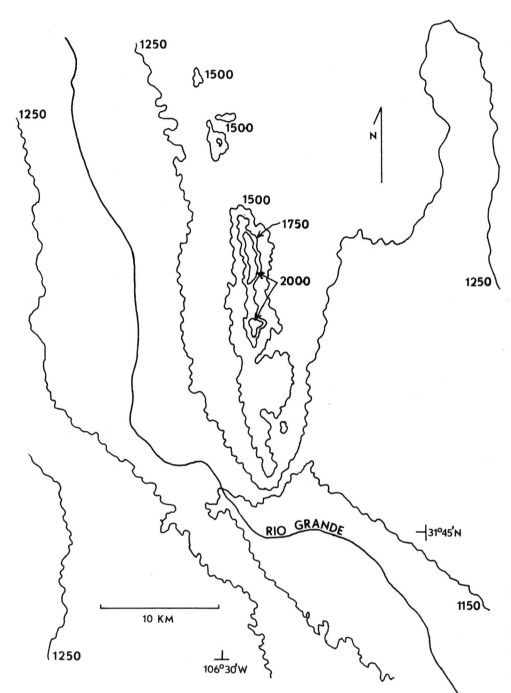

Figure 2.1. Topographic map of El Paso area. Altitudes
are in meters MSL.

The second case is essentially the condition for *dust storms* when the pressure gradient is intense. The mountain barrier sets up standing waves in the atmospheric flows which, if they develop sufficiently, extend downward to the surface and stir the boundary layer in a catastrophic fashion. Impinging of these lee waves on the desert floor results locally in large amounts of loose materials being picked up by the winds and then being transported upward by the intense eddy diffusion associated with such disturbances in the orderly progress of wave motions. This situation is most common in the flatlands to the east of the mountains since winds are predominantly from the west, although on occasion the situation may be reversed.

In addition to these extreme cases, the complex terrain of the El Paso area introduces a variety of mesoscale effects through local circulation systems produced by differential heating and cooling of local areas as a result of altitude, surface conditions and degree of pollution of the lower atmospheric layers. These local valley-type winds may produce altogether different types of air masses in different parts of the city from these natural processes as opposed to the technological pollution which man imposes on his environment.

The geography of the El Paso area can then be considered to have a most important impact on meteorological processes which form the structure of the atmosphere over the city. The complex static local variations which are to be expected from differing elevations and surface materials are greatly complicated by dynamic interactions between the general circulation and these orographic barriers to that flow.

ATMOSPHERIC OBSERVATIONS IN THE EL PASO AREA

Routine records of meteorological observations were begun at El Paso in August 1850. These records were maintained by the Army at various locations in the Paso del Norte area until this work was turned over to the Weather Bureau on 1 July 1891. The Weather Bureau was a part of the Department of Agriculture during this early period, and the El Paso station operated during this period as a climatological and six-hourly map observational point, which was located at various sites in the downtown area.

Meteorological requirements changed with the coming of air transportation to El Paso, so the Weather Bureau established an hourly observational station at the airport (see point A, Figures 1.1 and 3.1) on 20 November 1931. The downtown station was closed on 1 July 1939, and the Weather Bureau was transferred to the Department of Commerce on 30 June 1940. Pilot balloon observations of the winds aloft began at the airport station on 1 April 1932, and radiosonde observations were begun on 12 July 1939. Airplane soundings of the free atmosphere were initiated on 1 July 1935, at El Paso and carried out until balloon-borne radiosondes were initiated in 1939.

An intensive meteorological observational program was conducted at Biggs Field (see point B, Figures 1.1 and 3.1) in association with Air Force operations there during the 1950's and early 1960's. This station was initiated in April of 1941, and, for a period during the height of Air Force activity (1959-1966) at that location, the weather detachment included complete surface and special observational units along with a forecast unit. This meteorological station was closed in April 1966 with the termination of intensive flight operations at Biggs Field.

Meteorological observations were initiated at White Sands Missile Range in 1945 with the beginning of rocket testing at that site. Surface observations of a variety of types have been carried out at numerous sites on the range, with a complete observational and forecast station located at the headquarters area at the south end of the range (see point C, Figure 1.1). This locale also includes a radiosonde station (with five other intermittent radiosonde stations on the range), and in 1958 a rocketsonde station for sounding of the free atmosphere routinely to altitudes as high as 80 km was added.

Current plans call for activation in 1971 of a meteor trail radar at White Sands Missile Range which will provide wind observations to 110 km, making it possible to provide wind profiles on a daily basis from the surface to 110 km over the El Paso area. Additional special atmospheric observational equipments employed in the headquarters area

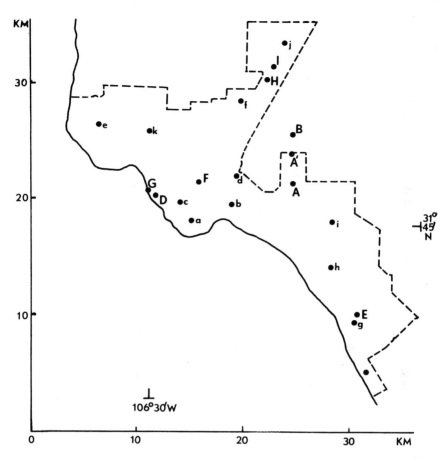

Figure 3.1. Location of El Paso meteorological observation points providing data used in this study. The solid (Rio Grande) and dashed lines indicate the city limits. Exact locations are given in Table 3.1.

include an ionospheric sounding station (C3 modified down to 250 kHz) which has operated continuously since 1947 and two magnetometer stations which were initiated in 1969.

These major centers for meteorological observations in the El Paso area are supplemented by numerous locations at which limited observational programs are conducted. The known observational points of this type are indicated in Figure 3.1., and an abbreviated listing of the observational program and period of record for these locations is presented on the following pages.

3.1. Surface Observations.

A considerable amount of data relative to various atmospheric parameters exists for the El Paso metropolitan

Table 3.1. Locations of special meteorological stations in
El Paso.

Name	Indicator	X	Y
Weather Bureau Surface Station	A	24.1	21.6
Weather Bureau Radiosonde Station	A'	24.0	24.0
Biggs Field	B	24.4	25.6
American Smelting and Refining Co.	D	11.6	19.8
Ysleta, Bureau of Reclamation	E	30.4	10.2
KROD Transmitter	F	15.4	21.2
American Dam	G	11.2	20.1
Chris H. Roach, 9120 Mt. Shasta	H	21.5	29.8
C. B. Dane, 9505 Verbana	I	22.5	31.1
El Paso Fire Department Stations:			
600 E. Overland	a	15.2	17.1
4240 Alameda	b	19.5	18.7
301 Robinson	c	13.6	19.2
3801 Ft. Blvd	d	18.8	22.5
115 Shorty Lane	e	6.3	25.8
8485 Sheridan	f	19.5	28.2
8803 Alameda	g	29.1	10.2
7901 San Jose Rd.	h	27.2	14.2
2405 McRae	i	27.4	18.6
10000 Dyer	j	22.6	32.9
6500 N. Mesa	k	10.8	25.1

area in the form of intermittent records kept by individuals
and firms. While it is most difficult to locate all of this
type of record, many of those data available for El Paso
are indicated by the following listing. These stations are
located by street number, by their general mesonet grid
location on the map of Figure 3.1 and their mesonet grid
coordinates in Table 3.1. The general nature of these records is indicated in the following listing.

1. American Smelting and Refining Company (Point D): Wind
and temperature measurements have been made on the roof of
the Environmental Sciences Department office since 1967. A
50-foot tower activated in 1969 is located approximately .3
km south of the office with wind measurements at 50 feet,
while at the main stack wind is measured at 175 feet and
temperature is measured a 6, 175, 500 and 800 feet. A rain
gage has been operated in the headquarters area since 1941.

2. Ysleta (Point E): Precipitation, wind and temperature
measurements have been made at this site since 1940. The
station is operated by the Bureau of Reclamation and is
located at 31°42'N, 106°19'W and 3670 feet altitude.

3. KROD Radio and TV Transmitter (Point F): Accumulated
precipitation and ambient temperature have been recorded
daily at approximately 0700 MST since 1955.

4. American Dam (Point G): Precipitation measurements have
been made at this location since 1938 by the International

Boundary and Water Commission. The data are published in:
Flow of the Rio Grande and Related Data, Water Bulletin,
Nos. 29-38, International Boundary and Water Commission,
United States and Mexico. The rain gage is located at
31°47'N, 106°32'W and 3730 feet altitude.
5. Chris H. Roach, 9120 Mt. Shasta (Point H): Daily accumulated precipitation records have been kept by Mr. Roach since
August 1957.
6. C. B. Dane, 9505 Verbena (Point I): Daily accumulated
precipitation records have been kept by Mr. Dane since
February 1963.
7. El Paso City Fire Department Stations (Points a-k):
Records of storm rainfalls have been logged, and in many
cases daily accumulated records have been kept.

3.2 Balloon Observations

Radiosonde observations of the free atmosphere using
balloons for transport of sensors have been accomplished twice
daily by the Weather Bureau (Station A', Figure 3.1. and
Table 3.2) at El Paso since 1939. The current observational
schedule is for release at 0500 and 1700 MST. Prior to
1 June 1957, the release times were 0800 and 2000 MST. Per-
formance of these balloon systems has improved steadily,
with current burst altitudes averaging approximately 30 km
(near 100,000 feet). These data have been used to describe
the tropospheric circulation as presented in Figures 4.9-4.15.
Radiosonde observations are carried out at White Sands
Missile Range (Station C, Figure 1.1. and Table 3.2) on an
intermittent basis. These observations have been accomplished since 1947, with a recent average of 350 observations per month. A portion of these observations are made
at five other locations (see Table 3.2) on White Sands
Missile Range which are operated when required by the rocket
test program.
Radiosonde temperature measurements are obtained through
telemetry of the resistance of a semiconducting rod thermistor which is exposed to the ambient air. Water vapor
concentration is measured with another electrical element
where the resistance is a function of the amount of moisture
absorbed on a special film. Pressure is measured by an
aneroid cell system at lower altitudes, while a hypsometric
technique is substituted at altitudes above about 20 km.
From these temperature, humidity and pressure data,
the heights with time of the sensing systems are calculated,
using the hydrostatic equation, and these data are combined
with angular measures of balloon position to evaluate drift
of the balloon with time as it rises. These data are then
used to determine the horizontal winds. Accuracies of these
radiosonde data have been estimated to be ± .7°C for temperature, ± 5% for humidity, ± 1 mb for pressure and ± 6
m/s for wind at 60 k feet under strong wind conditions by
the Meteorological Working Group of the Inter-Range Instrumentation Group (Meteorological Equipment Data Accuracies,

Table 3.2.   Location of balloon-borne radiosonde stations
in the El Paso area.

Station	Latitude	Longitude	Altitude(m)
A', Weather Station	31°48'N	106°24'W	1195
C, White Sands Missile Range	32°24'N	106°22'W	1216
C_1, Small Missile Range	32°28'N	106°25'W	1219
C_2, Apache Site	32°38'N	106°24'W	1206
C_3, Stallion Site	33°48'N	106°40'W	1506
C_4, Jallen Site	33°11'N	106°29'W	1235
C_5, Holloman	32°51'N	106°5'W	1247

IRIG Document 110-64, Defense Documentation Center, Cameron
Station, Alexandria, Virginia 22314).

3.3   Rocket Observations

Synoptic rocket exploration of the upper atmosphere
had its genesis at White Sands Missile Range in July 1958
with the firing of four Loki II meteorological rockets.   The
need for data in the 25-65 km altitude range was great in
the rocket test program at White Sands Missile Range, and
the Loki and Arcas meteorological rocket systems provided
the necessary vehicles for acquisition of these data.   The
White Sands Missile Range meteorological rocket launch site
(Point C', Figure 1.1.) became a primary station in the
cooperative Meteorological Rocket Network (MRN) and has served
as the testing ground for most of the techniques which have
produced synoptic meteorological data in the stratospheric
circulation region.

During the formative years, the announced schedule of
firings called for noontime launchings each Monday, Wednes-
day and Friday.   At the White Sands Missile Range station
special tests, research firings and operational requirements
quickly enhanced the actual launch rate to an almost daily
acquisition of data, and in July 1965 the announced MRN
schedule was changed to each work day year around.   The
annual firings accomplished to the end of 1969 at White
Sands Missile Range are presented in Figure 3.2.

The Arcas and Loki meteorological rocket systems simply
serve to transport meteorological sensors to the top of the
layer to be measured.   At peak altitude the rocket expulsion
system separates the sensors from the rocket, and the measurements are made as the sensors descend.   In all cases
radar track of a drag system (parachute, ballute, sphere,
etc.) provides a measure of the vertical profile of the
horizontal winds.   The characteristics of these parachute
type sensors have been considered by Murrow (1969), while
falling sphere techniques were analyzed by Wright (1969),
and chaff wind sensors have been discussed by Beyers (1969).
Dispersion of chaff clouds usually limits the altitude range
over which they can be used, so most MRN wind observations

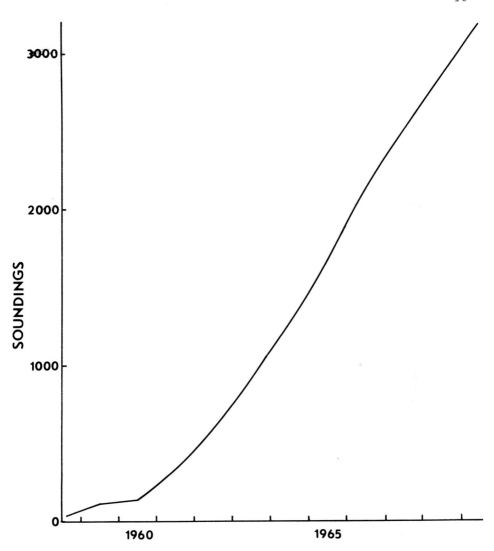

Figure 3.2. Annual total meteorological rocket firings
accomplished by the Atmospheric Sciences Laboratory at
White Sands Missile Range.

are made with parachute systems. There are instances, such
as at high altitudes, where chaff can do the job best.
While spheres have certain advantages over parachutes for
wind measurements, the parachutes are generally used in the
MRN because they provide a desirable platform for telemetry
systems and other sensor exposures. Wind date accuracies
of ±5 m/s to 50 km and ±10 m/s to 65 km altitude are obtained with the parachute and sphere systems.
      The primary sensor used on rocketsondes is a 10 mil

diameter semiconducting bead thermistor (Ballard, 1967).
The electrical resistance of this sensor is telemetered
over a 1680 MHz transmitter to a ground station for record-
ing. Data from this system are expected to be accurate to
±2°C from 25 to 55 km altitude and ±5°C to 65 km altitude.
A most important aspect of the temperature measurement is
the exposure of the bead. While in the rocket the bead is
at the top, but when on the parachute it is oriented downward with the bead the first part of the system in the un-
modified environment.

Another sensing system employed on the Arcas is an
ozone concentration detector (Randhawa, 1969a). This
sensor is based on the chemiluminescent technique in which
a known quantity of ambient air is passed across the sensor
and a photomultiplier is used to monitor the amount of light
released as the ozone is burned. These data are then telemetered to the ground and converted to ozone concentration.
Ozone concentration over the altitude range from 10 to 65
km is observed to an accuracy of ±10% with this instrument.
A measurement of temperature is generally included in the
ozone package, yielding wind, temperature and ozone profiles
on those flights.

Of particular importance relative to these sensors is
the sensitivity which has been achieved. These small systems
produce response to 1/e of a step function in a 100 meter
altitude interval between 25 and 50 km and at 1 km intervals
above 50 km.

A special sensing system used on the Arcas rocket is
the cryogenic sampler (Ballard, 1970). This device is
operated at an internal temperature of approximately 25°K
while airborne by boiling liquid neon out of the system.
After deployment from the rocket onto the parachute, the
sampler is opened for a short interval. All ambient air
(most of which freezes at roughly 45°K) which enters the
sampler is trapped, either through sublimation or due to
the pressure gradient. After the sampler is closed, the
interior warms up and the collected air goes back into the
gaseous phase. The sampler is recovered, and the sample
is analyzed in a mass spectrometer.

The rocketsonde station at White Sands Missile Range
is providing daily profiles of atmospheric structure in
the stratospheric circulation for the El Paso area. Data
from the sensor systems discussed above have been used to
prepare the curves illustrating the structure of the stratospheric circulation which are presented in Figures 4.16-4.23.

3.4   REMOTE SENSING

Systematic observation of the meteorology of the thermosphere (that region above 80 km altitude) in the El Paso
area is just beginning. The initial observational system
which will begin operation in 1971 at White Sands Missile
Range is the Meteor Trail Radar (MTR) (Duff, 1971).
The MTR technique involves measurement of drifts of

electron trails generated in the 80-110 km region by incoming meteor particles. These drifts are interpreted to indicate the motions of the neutral winds of that region and thus provide a measure of the dynamic processes of the lower thermosphere. In addition, the MTR observations can, through certain assumptions, be used to evaluate the ambient density and thus provide additional information on thermospheric structure.

The White Sands Missile Range MTR operates near 30 MHz and delivers pulses of 50 kW intensity to the north and to the east 250 times per second in the search mode. Detection of a return which meets the specifications of a meteor trail return causes the radar to switch to a measuring mode of operation, and for the next one half second the radar samples the radial drift of the trail at a rate of 500 times per second. Thus, the data obtained are discrete east-west and north-south measurements obtained from only those meteor trails which happen to occur normal to the radar beam. Altitude resolution is expected to be better than 2 km.

Observations of the electron density as a function of height have been conducted at White Sands Missile Range since 1947. These data are obtained at fifteen-minute intervals with a C3 ionosonder modified to sample down to .25 MHz in frequency. These data are applicable for studies in communications and detection problems and for analysis of ionospheric modification efforts.

The ionosonder discussed above provides a picture of the electron density structure of the lower ionosphere as if it were a static region of the atmosphere. In fact, large electric currents flow in this region of the atmosphere, and the electrons which reflect the radio waves of the ionosonder are really in transit across the observational space with speeds in the km/sec range. These electric currents produce magnetic fields through dynamo actions, and these magnetic fields can be observed at the earth's surface as modifications of the permanent magnetic field of the earth.

3.5 AIR POLLUTION OBSERVATIONS

The American Smelting and Refining Company (ASARCO) has established a network for continuous sampling of sulfur dioxide (SO_2) emissions in the El Paso area. Conductivity-type samplers of special construction with telemetry to the Environmental Sciences Department office are used to control the emission of SO_2 from their tall stack so that concentrations in the metropolitan area will stay within acceptable limits.

The ASARCO SO_2 network consists of 18 stations. This SO_2 network represents the most comprehensive air pollution sampling network to be deployed in the El Paso area thus far. It also represents the first attempt to make industrial use of these techniques in adjusting El Paso's technology to a working basis with the metropolitan population.

The ASARCO sulfur dioxide sampling network is tied di-

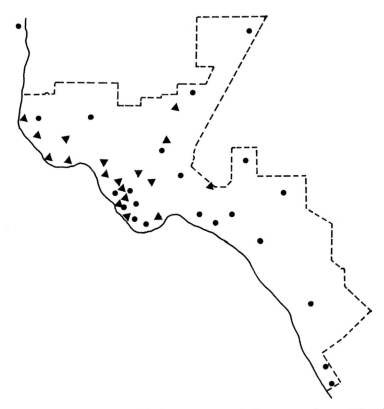

Figure 3.3. Stations which were used for an air pollution
study at El Paso in 1968 (Air Quality Survey, El Paso Metropolitan Area, October 15-November 15, 1968. Texas State
Health Department, Austin, Texas). ASARCO STATIONS are indicated by triangles.

rectly by telephone line to an IBM 1802 computer which
provides 15 second average values from the 18 sensors for
display at the control center. The computer is also programed to provide control personnel with wind and temperature data which are applicable to the dispersion problem.
     Measurements of air pollution parameters have been
made by the El Paso City-County Public Health Department.
The distribution of sampling stations used in the 1968 intensive survey of local air pollution characteristics is
illustrated by the dots in Figure 3.3. Some of the Health
Department's sampling equipment is mobile, and the majority
of the sites indicated in Figure 3.3 are used intermittently.

CLIMATOLOGY

No formal climatology of the El Paso area has been published, although there have been numerous articles dealing with various aspects of the climate of the region published in the literature. This chapter will be devoted to a preliminary look at the climatological structure which can be derived from the data available. While this material is considered to be informative, it clearly falls far short of the type of climatological information which should be readily available for mundane city planning and operational usage.

Principal problems with meteorological data in the El Paso area center around continuity of records and adequacy of instrumentation and exposures. Too great a reliance has necessarily been placed here on the data from a few select observational points, and thus this limited look at El Paso's climatology cannot present the complete story. It could even present a distorted picture. Even so, it will perform the valuable function of clarifying the meteorological data situation at El Paso as it currently exists.

The markedly inhomogeneous orographic features of the El Paso area assure a complexity in atmospheric structure. This is true during high wind conditions because of lee waves and turbulence induced into the flow by the mountains and results in low wind conditions as a result of drainage winds and other local circulations. The highly smoothed data which must make up a climatology must then be viewed in the light of the fact that a marked space and time variability exists in the El Paso area.

The atmosphere over El Paso is divided climatologically into three major circulation systems. They are the tropospheric circulation which operates in the 0-25 km altitude region, the stratospheric circulation which operates in the 25-80 km region and the thermospheric circulation which operates in the ionospheric region above 80 km. This dynamic division of the atmosphere into operating regions just happens to fall at about the same levels as the division by observational equipments. Balloon sounding systems, with their burst altitudes of approximately 30 km, do a rather complete job of sampling the tropospheric circulation. Meteorological rocketsondes obtain data in the 25-80 km altitude region, while electrical sounding systems provide the data source for the thermospheric region. After a look at the surface meteorological data, the above breakdown of atmospheric structure will be used in this treatment.

This approach to a climatology of the El Paso area is comparatively reasonable, as it would be for most midlatitude stations. However, at high or low latitudes, the above outlined breakdown of the neutral atmosphere circulation

into three well-defined systems may not be the most desirable or even sufficiently representative. The data are
already adequate to show that large variations occur in the
levels of the stratonull surface (separating the tropospheric and stratospheric circulation systems) and the stratopause surface (the temperature maximum near 50 km altitude). It seems likely that these dynamic separations will
become less impressive to us as the available data increase.
        While the atmosphere over El Paso has been explored in
greater detail than has any other location on the globe,
the upper reaches of the atmosphere remain a considerable
question, even over El Paso. The data which are available
generally concern the electrical structure, and it is only
with certain assumptions that these data are translated into
those parameters of particular interest to the meteorologist.
The Atmospheric Sciences Laboratory at White Sands Missile
Range is currently engaged in extending its observational
capability into the lower thermosphere. It will soon be
possible to look at that region with the same assurance with
which we now inspect the tropospheric and stratospheric
regions. Until then, it will be necessary to infer the
actual structure from a small amount of data and indirect
inferences from the circulations observed below that region.

4.1. Boundary Layer Climatology

        Systematic surface meteorological observations were
initiated at El Paso in 1877. Various locations were used
by the United States Army Signal Corps and the United States
Weather Bureau, with emphasis on what is now the downtown
area. In 1942 the principal observation point was moved to
the El Paso International Airport, its current location.
The station is situated at 106°24' West longitude, 31°48'
North latitude at an elevation on 1194.5 meters (3918 feet)
MSL in the metropolitan position indicated in Figure 1.1.
        The climate and weather of El Paso are currently com-
monly described by the data acquired during the past 28 years
at this latter location. Clearly, this approach assures a
highly simplified picture of the very heterogenous atmospheric structure which is imposed on the local atmosphere
by the complex local terrain outlined in Chapter 2. While
a small amount of information exists to illustrate this
heterogeneity in local atmospheric structure (see Chapter 3),
a reasonable understanding of the climatology and meteorology must await development of the mesoscale meteorological
observational network which will be required to deal with
local air pollution and weather modification control problems.
        In addition to the brief summary of El Paso surface
climatology presented below, the interested reader is referred to extensive studies of the climatology of White Sands
Missile Range prepared by Taft and Hoidale (1968, 1969a, b,
c, d, e, f), and extensive 500' tower data (Glass, 1964).
        The annual trend of surface temperatures at El Paso is
illustrated by the curves of Figure 4.1. These data were

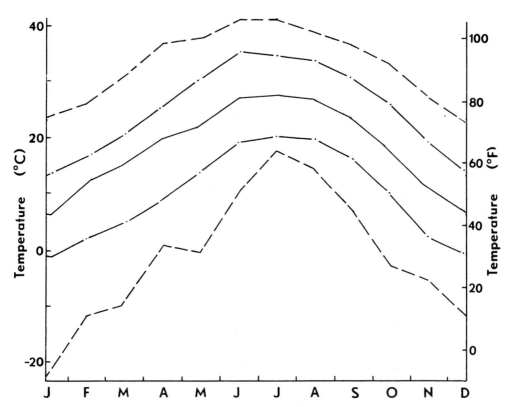

Figure 4.1. Normal monthly means (solid curve), mean maximums and minimums (dash-dot curves) and extreme (dashed
curves) temperatures for El Paso (A) published by the Environmental Science Services Administration, 1968.

published by the Environmental Science Services Administration in their Local Climatological Data report for El Paso
in 1968. Under clear skies and low wind conditions these
minimum temperature data will be significantly cooler than
the temperatures which prevail in the foothills of the
Franklin Mountains, and they will be even less representative of the highest mountain elevations. On the other hand,
radiation and drainage processes will produce the coldest
temperatures in the valley and thus make these minimums too
warm to be representative of those regions. The mean minimums and extreme minimums of Figure 4.1. are thus representative of the flat lands in the El Paso area, with local
minimum temperatures commonly varying under these conditions
over a range of tens of degrees Celsius as a function of
local elevation.

     Maximum temperatures reach highest values near summer
solstice time. A significant drop in maximum temperature
occurs early in July as a result of increased cloudiness
(see Figure 4.2), humidity (see Figure 4.3) and precipita-

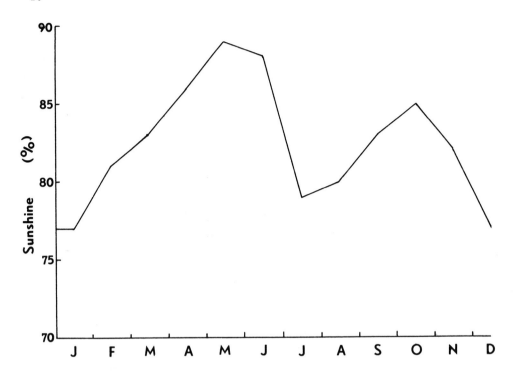

Figure 4.2. Monthly mean percent of possible sunshine at
El Paso as published by the Environmental Science Services
Administration, 1968.

tion (See Figure 4.4). Moist air from the Gulf of Mexico
pushes into the El Paso area during mid-July with associated
thunderstorm activity which serves to provide a sharp break
in the desert summer heat.
     One of the important characteristics of El Paso weather
is the bright sunshine which is indicated by the data presented in Figure 4.2. High clouds are frequently present,
so the skies are not generally completely clear. They do
little to obscure the sun, however, so a large amount of
surface solar insulation is characteristic of the El Paso
area. With little in the way of vegetation to shield the
ground, the surface becomes very hot during spring and early
summer afternoons. Superadiabatic lapse rates (temperature
decreases faster with altitude than 1°C per 100 meters:
5.6°F per 1000 feet) are common, and the presence of many
dust devils in the area is to be expected at midafternoon
in summer.
     The dominant dry dusty mode of mechanically transport-
ing heat vertically changes to a moisture-controlled buoyancy process in mid-July, and the cumulus and cumulonimbus
clouds, along with their associated cloud systems, shield
the ground from the sun's heat. This deepening of the
mixing layer from a few thousand feet in the winter season

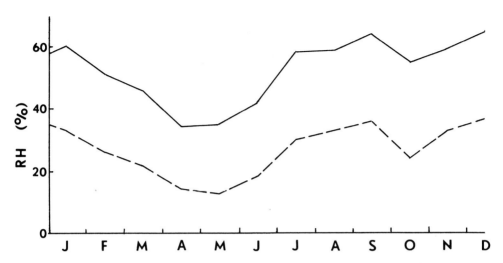

Figure 4.3. Normal monthly relative humidity values for El Paso at 0500 MST (solid curve) and 1700 MST (dashed curve) as published by the Environmental Science Services Administration, 1968.

to more than ten thousand feet in summer, coupled with the drastic redistribution of radiant heating which results from increased cloud cover, provides a most pleasant late summer for El Paso.

Monthly mean relative humidity values illustrated in Figure 4.3 follow a course prescribed by the diurnal and annual temperature variations and the dominant air mass situation. Most of the year the source region for El Paso's atmosphere is located to the west. This means that Pacific air must cross higher mountainous territory before arriving in the El Paso area with an attendant loss of moisture through western mountain slope precipitation. In addition, the dry surface of most of the region for several hundred kilometers to the west imposes strong drying action, particularly in the more stagnant situations.

Influx of moist Gulf of Mexico air into the El Paso area for the late summer and early fall period is clearly illustrated by the strong increase in relative humidity in July, August and September. This shift in air mass is precipitated by a westward development of the permanent Bermuda high-pressure region into the south central region of the United States as summer progresses. Circulation around this high-pressure cell brings air from the Gulf up the gentle slope of the West Texas plains. Progress of this westward flux of moist air is daily evidenced in June and early July by afternoon thundershower activity, particularly as the new air mass approaches a blocking mountain range such as the Sacramentos.

The Sacramento Mountains form a north-south-oriented barrier to this westward push of moist air approximately

40 miles east of El Paso, and an afternoon row of thunderstorms forms along that line for several days before the
moist air crosses the barrier. This latter maneuver is accomplished principally by means of an end run to the south
of the Sacramentos and up the Rio Grande Valley. During
this waiting period from mid-June to mid-July, El Paso has
its hottest weather, with daily maximums frequently in the
100°F (almost 40°C) range.
     Precipitation of water in El Paso is principally in
the form of rain showers with the main source from thunderstorms in late summer and early fall. The annual distribution of monthly precipitation means is presented in Figure
4.4. The annual total has a normal value of 20.04 centimeters
(7.89 inches) which is inadequate for most plants, at least
with the very asymmetric distribution in which it is received,
so that only the hardier desert plants populate the area.
     It may be inferred from the extreme data in Figure 4.4
that the rainfall is characteristically distributed in relatively large packages at irregular intervals. These rain-
storms are in the form of isolated thunderstorms in late
summer but on occasion represent the results of a more general
storm system in the upper atmosphere over the southwestern
United States.
     Reasonably comparable precipitation data are available
for the various Weather Bureau station locations in El Paso
over the past 40 years. These data are illustrated for the
annual totals in Figure 4.5. While the annual totals stay
fairly close to the established normal of 20.04 centimeters
(7.89 inches) of precipitation for the year, there are occasional large deviations, particularly toward greater amounts.
Principal features of these moist years are markedly greater
monthly amounts of precipitation during the normally dry
spring and relatively dry fall periods. Precipitation during
these periods in the El Paso area generally results from
special tropospheric circulation systems associated with a
Southwest Cutoff Low-pressure system centered in the global
500 mb circulation pattern. Circulation around this low-
pressure center, after the cutoff low has separated from
the generally global circulation, results in a southerly
inflow of moisture from the Pacific over a trajectory which
does not include all of the usual intense uplift and drying
actions.
     The cutoff lows remain in the region to the northwest
of El Paso for periods of several days on occasion, and when
they do move out they move to the northeast, generally intensifying as they go. The impact on the annual distribution of this mode of precipitation in the El Paso area is
illustrated by the data presented in Figure 4.6. In both
cases the spring and fall season monthly totals are several
times greater than normal (see Figure 4.4), with the summer thunderstorm contributions possibly even less than normal.
     Precipitation data are currently inadequate for a very
necessary comprehensive analysis of local isopleths in the

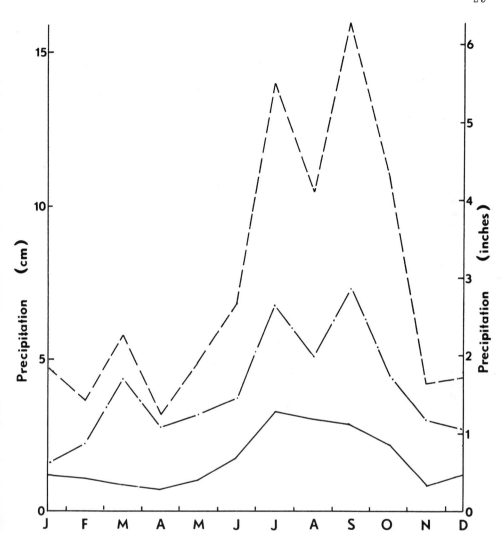

Figure 4.4. Normal monthly means (solid curve), maximum monthly totals (dash-dot) and twenty-four-hour extremes of precipitation at El Paso as published by the Environmental Science Services Administration, 1968.

El Paso area. Annual mean data for the past ten years for five locations scattered over the metropolitan area are presented in Figure 4.7. These annual totals give some indication that the flatlands on the east side of the Franklins receive more precipitation than stations on the mountains or to the west of the mountains. This result could in part be produced by a bias in the measuring systems, since rain gages in general do not perform well in windy conditions, particularly when vertical motions are involved. It is in-

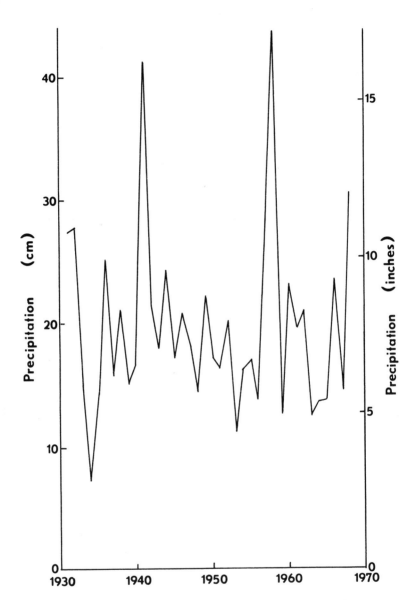

Figure 4.5. Annual precipitation for station (A) measured
by the Weather Bureau.

teresting to note that the Mount Franklin gage (KROD) has
the least desirable orographic situation for such a measurement. The eastern stations have similar locations, with
the only clear-cut difference being that the Weather Bureau
station is located in the downwind path of westerly flow
through the pass. These data thus provide some small indication that western, ridge line and pass locations receive

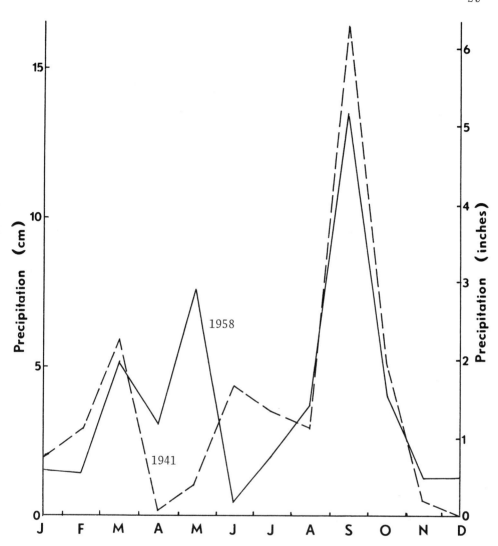

Figure 4.6. Monthly precipitation measured at the Weather Bureau station (A) total for the years 1941 and 1958 when the annual totals were abnormally high.

less precipitation than the remainder of the El Paso region.
      Of utmost importance to the weather situation in El Paso is the wind speed. Monthly mean and hourly extreme values of this parameter are illustrated in Figure 4.8. The general arid conditions result in a top soil layer of fine particles which are picked up by eddy motions of strong winds through saltation processes to form disagreeable dust clouds. Presence of the Franklin Mountains contributes significantly to this condition through inducing additional turbulence into the flow. Late winter and spring are the

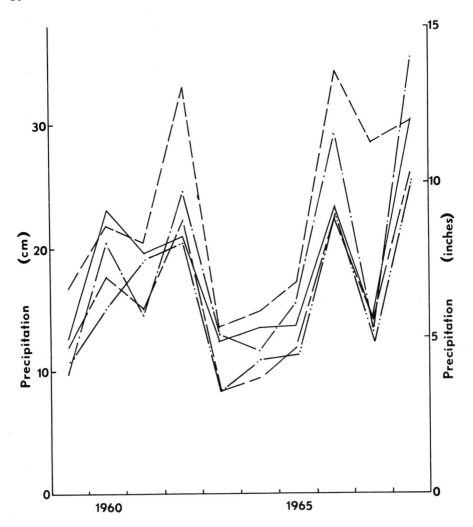

Figure 4.7.   Annual total precipitation for observation points
A (solid curve, Weather Bureau), B (dashed curve, Roach), C
(dash-dot curve, Ysleta) and F (dash-short curve, KROD Radio
and TV Station).   Point G (dash-dot-dot) is the American Dam
to the west of the Franklins.

seasons for dust loading of the atmosphere of El Paso.
Events of significance are very occasional even during the
period of maximum occurrence, and their duration is a few
hours on the average.

   The most important introductions of dust into the El
Paso atmosphere occur when mountain lee waves (discussed
in Section 4.2) develop in the troposphere as strong westerlies or easterlies flow across the topographic barrier of
the Franklin Mountains.   These waves reach great intensity

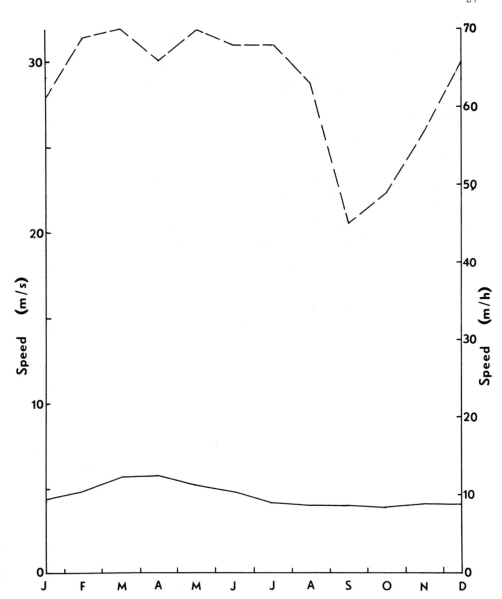

Figure 4.8. Normal mean hourly wind speed (solid curve) and fastest mile at El Paso as published by the Environmental Science Services Administration, 1968.

on occasion and literally pump dust up several thousand feet from the valley floor when the trough of the lee wave interacts with the ground. Prevailing winter and spring westerly components of the wind impose this severe condition on the Tularosa Basin region on the east side of the mountain range in general, although easterly winds associated

with cold high-pressure systems pushing into the central
plains brings similar weather to the west slopes of the
Franklins occasionally.

Dust is stirred up for short periods in the summer as
a result of high winds associated with thunderstorm activity. These phenomena are local to within an approximate 50
km diameter circle, and thus may engulf the entire city for
a short period of time. Such local dust clouds may not occur
if previous showers have dampened the region sufficiently
to prevent loosening of the small dust particles. These
events are isolated, and their approach is easily observed
so that the element of surprise inherent in many weather
phenomena is generally lacking.

At the airport station the strongest winds (fastest
mile) are from a westerly quadrant except in winter when
there is a northerly flow, veering even to the northeast
in January. The strongly dominating westerlies are in part
the result of a bias in location of the observation point,
since the station's downwind location for westerly winds
makes it a favored target for the strong winds of the lee
waves. If the station were on the west side of the mountain
range, the distribution would shift in favor of stronger
easterlies.

In general, winter winds move down the Rio Grande and
Tularosa valleys from the north and the summer circulation
is up the valleys from the south. The gap in the Franklin
Mountains introduces a complication into the above generali-
zations, with unimpeded flow along the course of the Rio
Grande through the pass relieving part of the easterly or
westerly oriented pressure gradients. This natural flushing
action of the atmosphere over El Paso assures that local
pollution accumulations will be expeditiously removed from
the metropolitan area throughout most of the year, although
in fall and winter the local air occasionally stagnates for
hours and sometimes for days, permitting significant accumulations of local polluting materials.

4.2. Tropospheric Circulation

The tropospheric circulation is a globally concentric
layer which is located in the lower 25 km of the atmosphere
and is powered primarily by solar radiant energy. Most of
the approximately 330 cal. m^{-2} sec^{-1} of the solar constant
of low latitudes which drives the tropospheric circulation
is applied to the earth's surface. Mixing processes dis-
tribute this heat through the lower atmosphere and upper
ocean, and lateral eddies then provide meridional transport of heat which serves to equalize the highly asymmetric hemispheric solar heat input. In addition to the in-
trinsic interest which a planetary circulation system ex-
cites, the tropospheric circulation system carries out the
important function of transporting water vapor and other
minor constituents from source regions into deprived areas.

In addition to the brief summary based on El Paso data

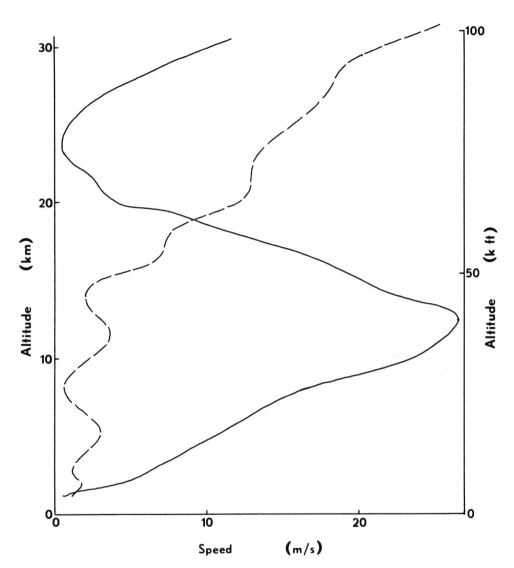

Figure 4.9. Mean monthly wind speed profiles for El Paso from the 0500 MST soundings obtained in January 1968 (solid curve) and July 1968 (dashed curve).

which is presented here, the interested reader is referred to climatographies prepared by Hoidale, Gee and Seagraves (1968a, b, c, d, and e, 1969a, b, and c) and to the White Sands Missile Range Reference Atmosphere (Part 1) prepared by the Meteorological Working Group of the Inter-Range Instrumentation Group.

Winter and summer vertical structures of the tropospheric circulation are well illustrated by the monthly

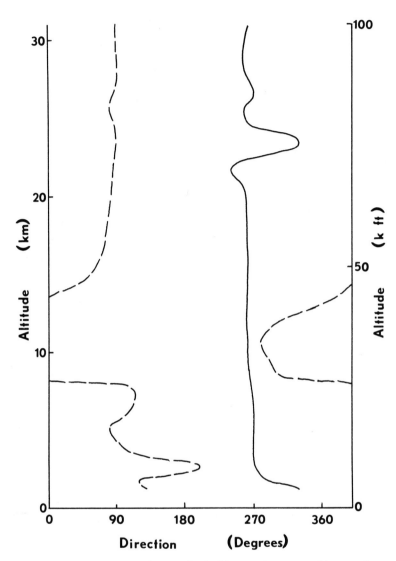

Figure 4.10. Mean monthly wind direction profiles for El Paso from the 0500 MST soundings obtained in January 1968 (solid curve) and July 1968 (dashed curve).

mean vertical wind speed profiles (Figure 4.9) and direction profiles (Figure 4.10) obtained by the Weather Bureau station at El Paso at 0500 MST during January 1968 (solid curve) and July 1968 (dashed curve). These data were obtained with the AN/GMD-1 balloon-borne sounding system which employs angular measurements of balloon positions and height determinations from temperature measurements and thermodynamic calculations. These data are characteristically

most accurate in early phases of the balloon flight and
deteriorate in accuracy as the balloon moves away from the
ground station (i.e., with altitude).

The wintertime curve presented here is highly characteristic of the general structure of the midlatitude tropospheric circulation. The tropospheric jetstream (Reiter,
1963) maximum in wind speed is normally located near the
tropopause, with strong positive wind gradients in the
troposphere of approximately 2×10^{-3} sec^{-1} and roughly equal
negative gradients in the lower stratosphere above. These
monthly average values are highly smoothed relative to the
individual observed profiles, with small-scale shear layers
exhibiting gradients an order of magnitude or more larger.

The wind speed minimum at 24 km in this vertical profile identifies the mean altitude of the 'stratonull' level
(Webb, 1965) over El Paso during January. The stratonull
surface represents the separation between the tropospheric
and stratospheric circulations and is characteristically a
pronounced minimum in the westerly winds of winter, occasionally even with a thin layer of easterly winds in the
stratonull region. The monthly mean maximum speed of jet-
stream winds of almost 27 m/s ($\sim$55 m/h) is frequently ex-
ceeded during the winter season, with speeds greater than
50 m/s ($\sim$100 m/h) from the west for intervals of several
days at a time (Reiter, 1963).

In summer, the average winds over El Paso are generally
easterly and rather weak as is illustrated by the data for
July 1968 (Figures 4.9 and 4.10). A slowly increasing wind
with height in most of the tropospheric circulation is indicated by these data. Above about 15 km altitude, the
gradient in wind speed more than doubles, and, after some
wavering in direction, the wind again steadies up on the
characteristic summer easterlies of the stratospheric monsoon.

The lower portions of both winter and summer vertical
wind speed profiles are strongly influenced by viscous interaction between the atmosphere and the earth's surface.
This friction term (**Fr**) is indicated by the last term of
the general equation of motion

$$\frac{d\mathbf{V}}{dt} = 2\mathbf{V} \times \boldsymbol{\omega} + \mathbf{g} - \frac{1}{\rho}\boldsymbol{\nabla}p + \mathbf{Fr}$$

where **V** is the wind velocity, $\boldsymbol{\omega}$ is the earth's rotation,
g is gravity, ρ is the density and p is pressure. Consequently
it is common to neglect frictional effects in the atmosphere
above the boundary layer through assumptions such as those
associated with the geostrophic wind (**V***) relation

$$f\mathbf{V}^* \times \mathbf{k} = \frac{1}{\rho} \boldsymbol{\nabla}_h p$$

where **k** is a unit vertical vector and the subscript h refers
to the horizontal plane. The presence of turbulent eddy
diffusion in the planetary boundary layer results in transport of momentum out of the circulation and thus serves to

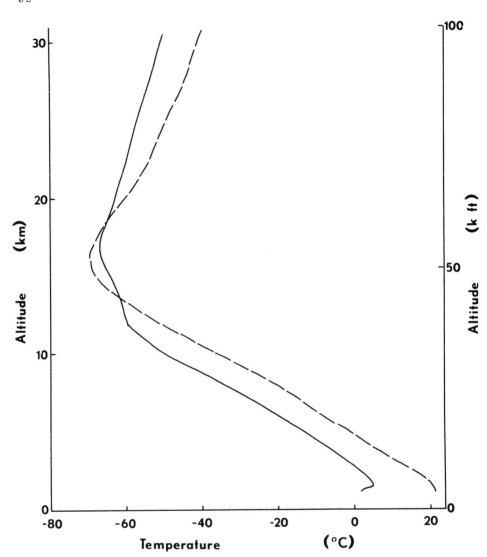

Figure 4.11. Mean monthly temperature structures for El
Paso from the 0500 MST soundings obtained in January 1968
(solid curve) and July 1968 (dashed curve).

slow tropospheric winds below equilibrium values. This
process causes the wind direction to veer with decreasing
height and tends to relax the pressure gradient through
transport of mass from high to low pressure.
     In addition to the solar radiant heating of the surface,
the strong eddy motions which are characteristic of the
troposphere exert a dominating influence on the thermal
structure of this lower portion of the atmosphere to estab-

lish the vertical temperature structure presented in Figure 4.11. Mechanical turbulence induced by strong local wind gradients and by topographic interferences with the surface flow forces the vertical structure of heat flow (F_H) indicated by the relation

$$F_H = - K_e \, \rho \, c_p \, (\gamma - \Gamma)$$

where K_e is the eddy transport coefficient, ρ is the density, c_p is the specific heat at constant pressure, γ is the ambient lapse rate and Γ is the adiabatic lapse rate. This mechanical eddy diffusion (ranging between 10^{-1} and 10^2 m^2 sec^{-1}) serves to cause heat to flow downward excessively in the lower stratosphere and troposphere, cooling the tropopause region and adding to solar heating of the lower layers. Thermal convective systems, on the other hand, transport heat upward, tending to reduce the vertical lapse rate of the troposphere. Mechanical eddy diffusion tends to to dominate the overall situation, resulting in a tropopause which is lower and warmer in winter when tropospheric mixing is weak and confined to lower levels and higher and colder in summer when mixing is at a maximum.

These early morning mean profiles illustrate the daily surface inversions which tend to form over the Pass of the North during portions of the year, particularly during the winter season. These very stable layers of the lower few hundred meters of the atmosphere reduce vertical transport and thus confine polluting materials to a thin surface layer of the atmosphere. It is when this situation is coupled with light winds that pollution conditions are at their worst in El Paso. Such conditions occur most frequently during fall and winter (see Figure 4.12).

Diurnal heating and cooling of the earth's surface by radiational processes produce a rather widely varying lower boundary on the vertical temperature structure of the atmosphere. El Paso's high altitude, clear skies and surface characteristics generally result in relatively unstable surface layers in the afternoon and very stable surface layers at night. Radiation losses during the night produce temperatures of the air near the surface which are significantly lower than those of the air aloft. This effect is a maximum when winds are light and just before sunrise.

These inversions (temperature increasing with height) serve to trap low-level pollutants, principally through reduction of vertical eddy diffusion intensity. On the other hand, pollutants released above the inversion do not penetrate efficiently downward for the same reason. A favorite approach by industry has been to build smoke stacks which will disperse the pollutants where winds and upward eddy transport will disperse the material without causing local problems. This system works well if the release is above the top of the inversion as long as the inversion persists. With morning heating of the surface, however,

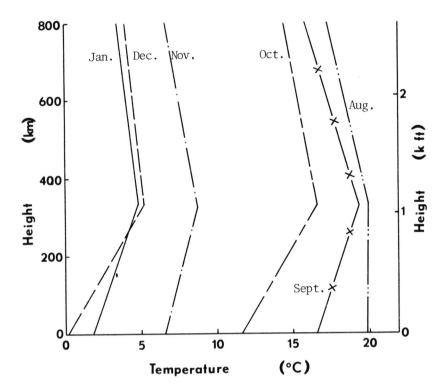

Figure 4.12. Mean monthly temperature profiles for 1968
for El Paso in the planetary boundary layer obtained from
Weather Bureau published values for the 0500 MST radiosonde.

such inversions usually do break up rapidly, and this
situation is drastically changed.

Breakup of a surface inversion results in large increases in the vertical eddy transport coefficient and
transports surface accumulations of pollutants rapidly
upward, thus resulting in a decided improvement in pollution
conditions. Breakup may occur in a matter of minutes, in
which case the second situation results in an increase in
pollutant concentration at the surface through downward
eddy transport. This process is called 'fumigation' and is
a serious problem for industry in that large amounts of
contaminant may be airborne above the inversion when breakup occurs.

Altitude of the early morning inversion layer and its
frequency of occurrence over El Paso have been inspected by
the Environmental Sciences Department of the American Smel-
ter and Refining Company. Radiosonde records obtained by
the Weather Bureau at the airport (more than 10 km east of
ASARC) at 0500 MST were analysed to obtain monthly mean
heights (solid curve) and the total number of occurrences
for the months indicated in Figure 4.13. The tops of the

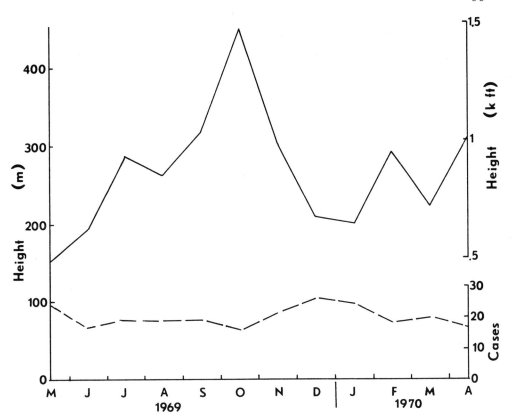

Figure 4.13. Monthly mean inversion top heights (solid curve) and frequency of occurrence (dashed curve) over El Paso obtained from the 0500 MST radiosondes at the Weather Bureau Airport Station. Data courtesy of the Environmental Science Department of the American Smelting and Refining Company.

surface inversions then average between 200 and 500 meters (600-1600 feet) above the surface. This means that the mountains extend above the inversion surface.

Synoptic meteorological conditions can exist under which the usual inversion breakup over El Paso is delayed or does not occur for a day or more at a time. These are critical air pollution conditions. In general this condition is prevented by surface heating or by increased low-level winds.

Solar heating of the earth's variable surface results in production of a very warm surface layer which is inhomogeneous in the horizontal plane. Buoyancy forces result in selective lifting of surface air over these warmer areas with an associated horizontal inflow of air into that area to be heated. The net result is development of organized convective systems with strong local upward motions which

are compensated for by weaker downward and lateral flows
over significantly larger areas. This convective mixing
process is of great importance in transport of atmospheric
impurities in the troposphere. It is very important to
note the difference between convective mixing and mechanical
eddy mixing center in the fact that convective processes
result in transport of heat upward, while mechanical eddy
mixing transports heat downward. Thus, convective mixing
exhibits a strong diurnal variation, carrying heat generated
by surface absorption of solar radiation upward during the
hot afternoon and generally becoming unimportant in the
early morning hours. Mechanical eddy mixing, on the other
hand, has its energy source in shears in the circulation.
While dirunal variations are to be expected in the circulation, they will diurnally be less intense and the diurnal
variation in mechanical eddy mixing should exhibit less intense variations than that illustrated by convective mixing.

Organized convective activity is a powerful mixing
process in the troposphere. It has its maximum development
in a tall thunderstorm cell which may reach to 25 km altitude (75,000 feet). In many cases the lifting condensation
level is not reached by these upward motions, and the latent
heat of condensation is not available to produce buoyancy
forces which will push on upward. The extent of this dry
convective activity is usually evaluated by calculating a
daily maximum mixing depth by noting the height of inter-
section of the ambient temperature profile and the potential
temperature specified by the surface maximum temperature
(Holzworth, 1964).

Monthly mean maximum mixing depths for El Paso are
presented by the solid curve in Figure 4.14. These data
illustrate the expected summer maximum, with the maximum
occurring somewhat early along with the early summer temperature maximum. Also presented by the dashed curve of
Figure 4.14. are calculated monthly mean values of the lifting condensation level for El Paso. In the hatched area,
cumulus-type clouds are probable during the moist summer
months since condensation will provide additional heat to
support organized convective activity before the top of
forced mixing is reached.

Mountain lee waves are special forms of gravity waves
with a principal characteristic of an upstream propagation
which is, in the mean, equal to the wind speed. That is,
lee waves are standing waves tied to some obstruction which
interferes with the flow. These waves are commonly noticed
downwind from mountain ridges, and they become particularly
obvious when they transport ambient air parcels to above
the lifting condensation level to form the very special
lenticular clouds along the crest of the wave. The droplets
which form these lenticular clouds have lifetimes measured
in seconds as they form along the forward edge of the cloud
in ascending motion and evaporate along the trailing edge
of the cloud as they are dried in descending motion.

Lee waves are also very obvious when the wave amplitude

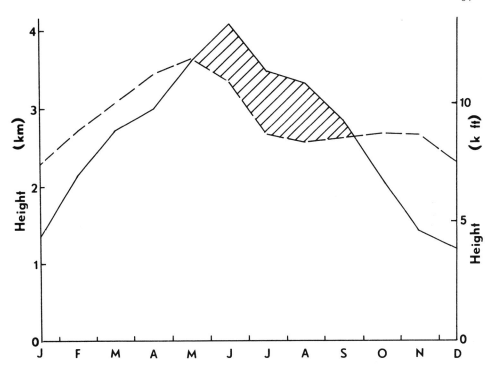

Figure 4.14. Mean monthly afternoon mixing depths for El Paso as calculated from 10 years of data through 1969 by E.S. Ethridge of the El Paso Weather Bureau Office.

becomes large enough to intersect the ground during the descending phase. In this case the trough of the wave is disrupted from its laminar wave motion as the air parcels strike the ground and are reflected. Soil particles are dislodged by this interaction and are carried aloft in various ways as the disturbed wave reacts with the atmospheric environment. Thus, intense lee waves may become visible at the crests and/or troughs and reveal the structure of the wave.

The intense lee waves discussed above occur only occasionally, even in mountainous regions. They generally require sharp ridge lines of several hundred meters elevation and wind speeds of several tens of meters per second. Optimum circumstances for formation of lee waves are when the ridge line is perpendicular to the wind flow, although the waves are observed when the flow is several degrees away from normal, and sometimes when the angle is a few tens of degrees.

In the El Paso area the waves are generally observed on the eastern side of the Franklin Mountains because strong flow situations dominately involve winds from the west. On occasion, however, easterly winds are strong enough to produce easily observable waves on the western side.

Lee waves frequently appear to the casual observer to
be quite stationary. Time lapse movies have demonstrated,
however, that these are a dynamic phenomenon and are in
constant motion, adjusting to fluctuations in the atmospheric
parameters which generate the waves. In a stable gas, buoyancy forces powered by a gravity field will restore a dis-
placed particle to its original potential level with a characteristic period of oscillation which is termed the Brunt-
Väisälä frequency

$$\omega = \frac{g}{T}(\Gamma - \gamma),$$

where g is the acceleration of gravity, Γ is the dry adiabatic lapse rate (1°C/100 meters), γ is the ambient lapse
rate and T is the temperature in degrees absolute. The
equation of motion of such a wave may be expressed as

$$\frac{\partial^2 w}{\partial z^2} + (L^2 - k^2)\, w = 0$$

where $L^2 = \frac{\omega^2}{U^2} - \frac{1}{U}\frac{\partial^2 U}{\partial z^2}$, w is the vertical velocity, k is
the horizontal wave number, U is the horizontal wind speed
in the undisturbed flow and z is the height.

Clearly, a certain stability of the air is required
for lee waves to develop. Otherwise, the flow will immediately degenerate into turbulence. On the other hand,
the amplitude of the wave will be greater in the less stable
situations. Also, a higher Brunt-Väisälä frequency, which
depends on greater instability, will provide a shorter wavelength and thus a more obvious wave. Fluctuations in lee
waves, both in amplitude and wavelength, will result in
radiation of wave energy of that frequency in the form of
gravity waves away from the source region into other regions
of the atmosphere. The evanescent mode (non-propagating)
retains most of the energy, but a few percent of the total
wave energy may be invested into these traveling waves.
These traveling waves are minor components of the energy
content of the lower atmosphere or the earth's surface
against which they may reflect, but the upward propagating
traveling waves may introduce significant amounts of new
energy into the upper atmosphere. Consideration of the
physics of propagation of such waves indicates that the
energy density of these waves will remain essentially constant in the lower atmosphere as a result of minor absorption and scattering losses which are characteristic of waves
with their gross dimensions. The normal rapid decrease in
density with height of the atmosphere then requires that
the amplitude of the wave increase with height in order to
maintain a constant energy density. At some altitude the
amplitude of the wave will become so great that accelerations will become excessive and the wave will 'break'.
The energy will then become converted through turbulence
into thermal energy of that region, and the wave will be
rapidly attenuated. This sink for gravity waves in the
lee wave frequency spectra becomes significant in the lower

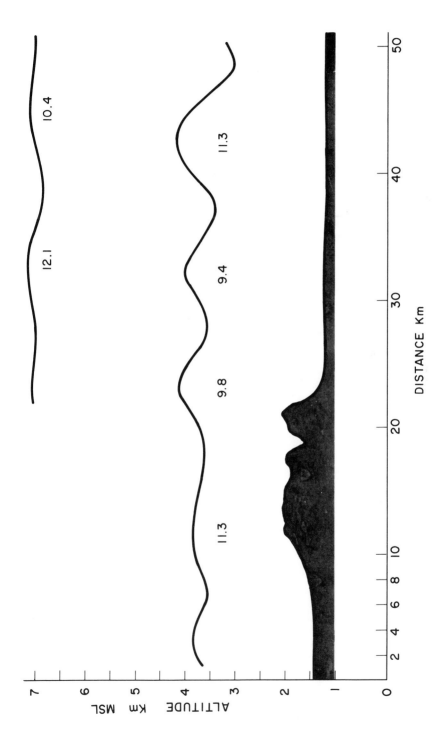

Figure 4.15. Streamlines of a lee wave observed by tracking of two super-pressure balloons over White Sands Missile Range near noontime on 6 May 1965.

ionosphere near the 100 km region.
      Lee Waves have been investigated at White Sands Missile
Range by Reynolds (1969), Reynolds et al. (1966, 1968),
Lamberth and Reynolds (1965), and Glass et al. (1968). By
radar track of superpressure balloons and telemetry of on-
board temperature and pressure measurements of the ambient
air, they have derived the general physical characteristics
of the lee waves of large intensity in the El Paso area.
Their results point out the great variability exhibited by
this phenomenon, and it is obvious that these waves have a
marked impact on the entire atmospheric structure in the
El Paso area.
      A typical measured cross section of a lee wave is
illustrated in Figure 4.15. Mean observed values of wavelength (λ) are approximately 18 km (12 miles) with values
as high as 37 km (23 miles). Since the Tularosa Bolson is
approximately forty miles across, there is room for several
waves before the before the next orographic barrier is en-
countered. Vertical amplitudes of 260 meters (853 feet)
were found on the average, with some observed waves of
greater than 800 meters (2624 feet) amplitude. Wind speeds
in these lee waves were typically several tens of meters
per second, which is also the order of the wave propagation
speed. Vertical velocities were measured as high as 8 m/s
(17.9 m/h), with an average of approximately 1 m/s.

4.3 Stratospheric Circulation

      The stratospheric circulation is defined as that glo-
bal monsoon wind system which operates in the middle atmosphere between approximately 25 and 80 km altitude. This
wind system results from a thermal field which is generated
principally by ozone absorption of solar ultraviolet energy,
a process which has its maximum effect in the 50 km region.
Principal data which are available for analysis of the
stratospheric circulation are wind, temperature and ozone
profiles obtained by the MRN. The physical structure of
this circulation system over the El Paso area, as it is
known today, is presented here.
      Subdivision of the atmosphere into specific units is
not without problems. While such divisions may be meaning-
ful in certain areas and for certain analyses, they generally
break down from the global point of view. This is because
the atmosphere actually operates as a whole, with the occasionally separated units yielding their identity at special locations and times to achieve the overall unity. The
stratospheric circulation is an excellent example of this
mode, appearing in some cases as a highly independent unit
and in other cases almost loosing its identity. Over El
Paso the stratospheric circulation operates more or less
independently, so it will be treated here in that sense.
      Approximately 3 cal m^{-2} sec^{-1} of solar energy in the
2000-3000Å radiation band is deposited in the stratospheric
region through absorption by ozone. The vertical distri-

bution of heat deposit is given to a first approximation
by the usual Chapman Layer (Chapman, 1951) assumptions
based on the relation

$$\frac{q}{q_{max}} = e^{1 - z - e^{-z}}$$

where z is measured from the level of maximum absorption.
Evaluation of this process is complicated by various un-
knowns, with the ozone concentration representing the
principal problem. Even with its faults, this technique
does give a general solution for the stratospheric heating
problem which is reasonable in the gross features.

Application of the Chapman layer concept generally
involves an assumed still, homogeneous atmosphere, and it
is in this respect that the theory evidences its greatest
shortcomings. Measurements of the wind and temperature
fields with the sensitive MRN sensors had already indicated
the presence of a high degree of inhomogeneity, and detailed
measurements of the ozone concentration field of the stratospheric circulation region have proved the ozonosphere to
be inhomogeneous in this respect also. Randhawa (1967,
1968, 1969b) has performed these measurements with a chemiluminescent rocketsonde in the 15-65 km altitude region
with results illustrated by the profiles presented in
Figure 4.16.

These ozone profiles indicate the general features of
ozone concentrations for the extreme seasons in the atmosphere over El Paso. These mixing ratio curves exhibit a
maximum of the order of 10^{-6} in the 20-25 km altitude region,
with a sharp drop in relative concentration near the tropopause to a relatively constant value of roughly 10^{-7} in the
tropospheric region. Large eddy diffusion transport coefficients which are characteristic of the troposphere result
in a strong flux of roughly 10^{11} ozone molecules per square
meter per second flowing downward onto the earth's surface.
The downward flux from the base of the ozonosphere must be
greater than this, however, since some of the ozone is lost
through oxidation and photochemical processes in the relatively dirty troposphere before it reaches the earth's surface.

The shape of the upper side of the ozonosphere is established by a photochemical equilibrium of sorts in which
the rates of formation and destruction and vertical eddy
transport achieve a local balance. At high altitudes the
ozone destruction rates are sufficient to prevent a buildup
of concentration, and eddy transport coefficients in the
mesosphere of the order of 10^{4} m^2 sec^{-1} provide a relatively
constant ozone mixing ratio except that the larger eddies
enforce a detailed structure as transient phenomena. Maximum lifetimes of atmospheric ozone molecules are found in
the 20-25 km altitude region, where the solar ultraviolet
has been absorbed to below ozone dissociation levels and
eddy transport processes are small enough to permit accumulation of this substance. This maximum in ozone relative

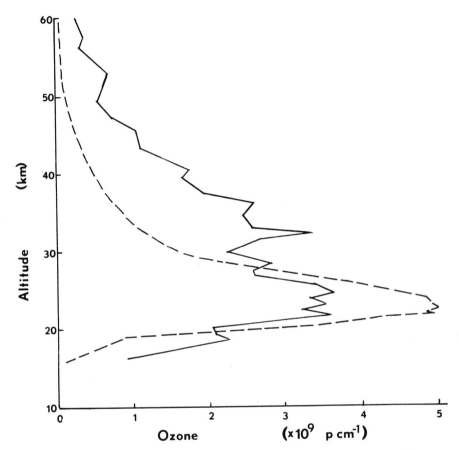

Figure 4.16. Ozone profiles observed over White Sands Missile Range on 18 February 1969 at 1000 MST (solid curve) and 17 July 1968 at 0428 MST (dashed curve).

concentration is a very stable feature of the stratospheric circulation.

Maximum integrated ozone concentration in the atmosphere is observed at high latitudes during the late winter and early spring period. This annual variation is produced by various transport processes which move ozone into the protection of the clean dark polar night. El Paso has little part in this global variation but stays relatively static in its ozone structure throughout the year.

With a known ozone profile, it is possible to estimate the heating which solar ultraviolet radiation will produce in the ozonosphere. Early attempts at this (Leovy, 1964) produced thermal structures of the general character which is observed in the El Paso area. A temperature of near surface values (0°C) was obtained for the stratopause region between 45-50 km, with lapse rates of roughly -.004°C m^{-1} above that level to form a stratospheric circulation which consists of a warm central layer bounded by cold lower and upper surfaces.

The first hint that this simplified picture of the thermal structure of the ozonosphere might be in serious error came in 1963 with meteorological rocketsonde observation of a diurnal temperature variation of the order of 15°C over the El Paso area. Since the theory had predicted a diurnal variation of only 4°C, this observation was of major importance. It was followed-up at White Sands Missile Range by a thorough study of the diurnal thermal and wind field of the stratospheric circulation by Beyers and Miers (1965) and Miers (1965), and has since been inspected at other times and latitudinal locations. The result of these studies is that there are large thermal variations on a daily basis near the stratopause, and this variable heating results in large pressure perturbations and associated variations in the wind field. These tidal-type variations are thus of major import to the physical structure of the stratospheric region as well as to other atmospheric regions, particularly in the case of those layers of the rarefied upper atmosphere above the mesopause.

Mean vertical profiles of the temperature structure of the stratospheric circulation region above El Paso for the winter and summer seasons are illustrated by the curves of Figure 4.17. These data were obtained through use of small meteorological rockets of the Arcas and Loki types and are published in the data reports of the MRN. They provide accurate mean seasonal profiles from 25 to 65 km altitude for the period of record, along with indication of the extremes which have been observed thus far.

These mean curves are smoothed by the data processing which has been applied to obtain these seasonal averages. The individual profiles which make up these mean profiles have a principal characteristic of considerable detailed structure. This variable vertical structure is illustrated by the cases depicted in Figure 4.18. Such small-scale features are almost surely formed by dynamic processes, with turbulent eddies and gravity waves or some combination of both being the most likely stirring mechanisms. Vertical motions induced by these phenomena will produce adiabatic temperature variations as well as inhomogeneities in ozone distribution which will then produce inhomogeneities in absorption of solar ultraviolet.

The thermal structure of the stratospheric circulation region in the El Paso area can then be considered to be well known for the period of observation in the 25-65 km altitude region. Consideration of the data presented in Figures 4.17. and 4.18. indicates that a degree of variability exists which, particularly for the winter storm period, is larger for the extremes of the mean profiles than is indicated by the individual profiles. This indicates that synoptic scale systems are at work in the stratospheric circulation, and it is well known from analysis of

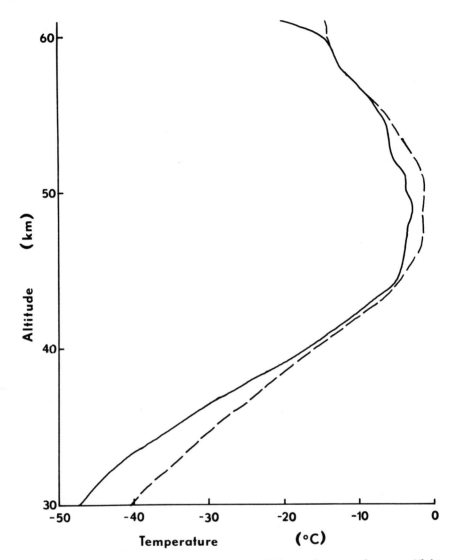

Figure 4.17. Mean temperature profiles observed over White
Sands Missile Range during December (solid curve) and July
(dashed curve) for the years 1961-1968 (Meteorological
Rocket Network Firings, Vol 5, Nos 7 & 12, World Data Center
A, Ashville, North Carolina).

the global MRN data that hemispheric and global scale synoptic systems do exist. These global systems have been observed
to produce temperature increases in winter high latitudes
at 35 km altitude of as much as 80°C in a one-week period
during the very special 'explosive warming' periods (Scherhag, 1952). These explosive warmings are occasional events

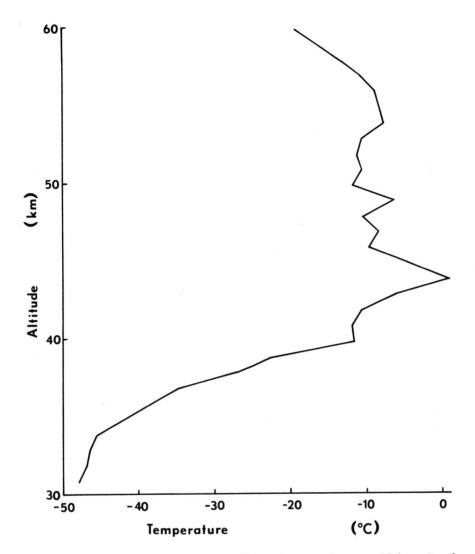

Figure 4.18. Temperature profile observed over White Sands Missile Range on 6 December 1968 at 0120 MST.

which occur during some Northern Hemisphere winters, invariably during the winter storm period.

The low-latitude location of El Paso tends to reduce the amplitude of temperature variations which occur in the stratosphere over that area. The nature of these variations is illustrated in Figure 4.19. The winter season variations of several degrees which occur with periods of several days result from oscillations of the North American trough which extends toward low latitudes. The axis of this trough oscillates eastward and westward from a central position over

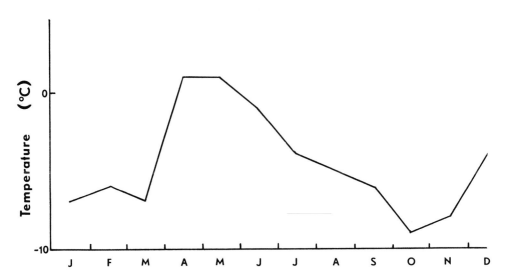

Figure 4.19. Monthly mean temperatures observed in the 45-
55 km altitude region over White Sands Missile Range during
1961-1968.

North America and induces first southwest and northwest winds
into the El Paso area with their resulting differing thermal
fields.

     The stratospheric circulation is motivated by meridional
gradients in the temperature field of the ozonosphere. This
driving force is indicated by the thermal wind ($\mathbf{V_t}$) relation
presented in the following equation.

$$\frac{\partial \mathbf{V_t}}{\partial h} \times \mathbf{k} = \frac{g}{2\omega \sin \phi T} \nabla T$$

Examination of the data from other MRN stations has shown
that the meridional temperature gradient exhibits an annual
variation over El Paso (as well as most of the hemisphere),
with strong negative values in winter and weaker positive
values in summer. This produces the grand monsoon of the
stratospheric circulation which consists of strong westerly
winds around a polar low-pressure vortex in winter and
easterly winds around a polar high-pressure system in summer.

     The character of the stratospheric monsoon over El Paso
is illustrated by the data presented in Figure 4.20. Mean
values of the zonal and meridional flows in the 45-55 km
altitude region (the SCI) have been averaged for the period
of record to obtain this mean structure of the annual stratospheric circulation for the El Paso region. These data
illustrate the fact that the winter season of the stratospheric circulation is an extended period of roughly seven
and one-half months and the summer season is an abbreviated
period of roughly four and one-half months, with the re-
versal times falling at the first of May and at fall equinox

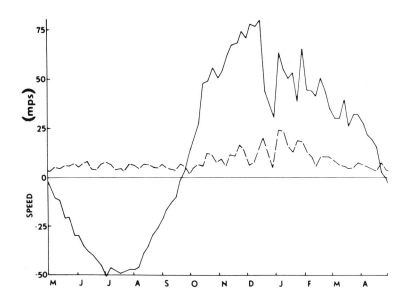

Figure 4.20. Annual variation of the zonal meridional Stratospheric Circulation Index (SCI) components for White Sands Missile Range, New Mexico. Five-day means are plotted, including all nearest noon soundings obtained through 1967. Positive values refer to winds from the west and south.

time. Also obvious in these data is the importance of the winter storm period to the stratospheric structure of the El Paso area.

Discovery of the stratopause thermal tides by Beyers and Miers introduced a new dimension to our knowledge of the structure of the upper atmosphere. The basic equation of motion for a perturbation of this type is given (after Lindzen, 1967) in the following equation

$$\frac{d}{d\mu}(\frac{1-\mu^2}{f^2-\mu^2}\frac{d\Phi_n}{d\mu}) - \frac{1}{f^2-\mu^2}(\frac{s}{f}\frac{f^2+\mu^2}{f^2-\mu^2}+\frac{s^2}{1-\mu^2})\Phi_n + \frac{4a^2\Omega^2}{gh_n}\Phi_n = 0.$$

The solar thermal heating function illustrated in Figure 4.21 has the distinctive properties of a range of roughly 15°C with maximum at 1400 LST and minimum near sunrise. This heating produces a global heat wave at the stratopause region which has the character illustrated in Figure 4.22. The low-frequency portion of the equation is illustrated by the arrows, showing a strong divergence of the flow in equatorial regions and a strong convergence of the flow in high latitudes.

The zonal phase of this thermal tidal circulation is markedly different for the summer and winter hemispheres. In winter the tidal circulation falls short of the pole and the zonal winds are westerly. In the summer hemisphere

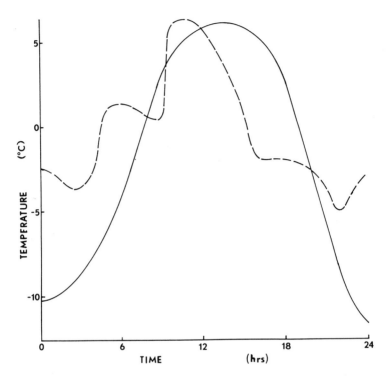

Figure 4.21. Diurnal temperature variations of the 45-55 km altitude region for White Sands Missile Range (solid curve) and Ascension Island (dashed curve). These curves were taken from Beyers, Miers and Reed (1966) and Beyers and Miers (1968).

the high-pressure ridge extends across the polar region, and the zonal winds become easterly in the high-latitude nighttime sky. This 'stratospheric tidal jet' of summer high latitudes is clearly instrumental in formation of the very special noctilucent clouds of that region.

In addition to the synoptic-type circulation systems which solar thermal heating of the ozonosphere impresses on the stratospheric circulation, there are traveling waves radiated away from this dynamic generator into other regions of the atmosphere and absorbed. These traveling waves are of little consequence in the lower atmosphere because the increased density of the air results in a reduced amplitude of the wave. Upward propagation, on the other hand, may be very important since the reduced ambient density causes the wave amplitude to increase until at some level the wave will become unstable and dissipate into turbulence and heat. This region of dissipation of gravity waves, whether from the diurnal tides of the stratopause or from lee waves or other sources, is generally found in the lower thermosphere in the 100 km region and above. Energy radiated by these waves

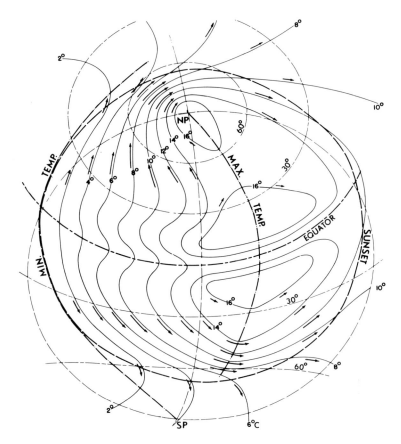

Figure 4.22. Model diurnal temperature and wind fields of
the 45-55 km stratopause region at the Northern Hemisphere
summer solstice. The map is an equidistant projection centered at 103°W and 42°N (Courtesy, Reviews of Geophysics,
Webb, 1966b, Figure 6).

out of the stratospheric circulation has been estimated by
Lindzen (1967) to be approximately twenty percent of the
input thermal tidal energy of the stratopause region.
    The structure of the stratospheric circulation is es-
tablished by the structure of the motivating forces, which
are principally density differences established by the thermal structure, and the structure of sources and sinks for the
kinetic energy of the circulation. The stratopause level,
defined as the temperature maximum of the ozonosphere, is
an indicator of the physical processes which determine the
structure of the stratospheric circulation. As is indicated by the thermal wind and the geostrophic wind in the previous equations the density decrease with altitude produces
stronger horizontal winds with increasing altitude of the
level of these winds. While most of the stratospheric cir-

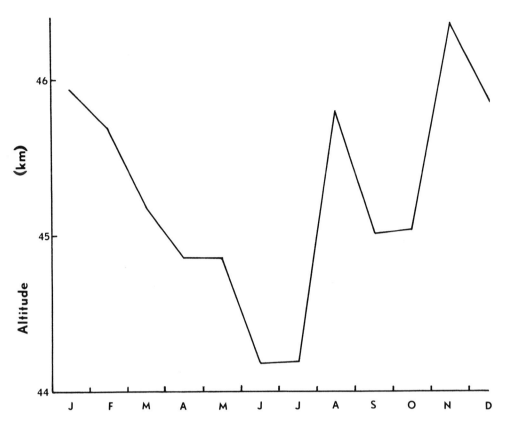

Figure 4.23. Mean monthly stratopause altitude for White
Sands Missile Range for the years 1961-8.

culation variability occurs at high latitudes, variations
of the stratopause are observed at El Paso latitudes. The
altitude variation which has been observed thus far in the
MRN data are illustrated by the curve in Figure 4.23.

These monthly mean altitudes of the stratopause level
show the stratopause to be at maximum altitudes near 46 km
during the winter season and minimum altitudes near 44 km
in summer. This overall annual variation is in general
agreement with that expected on theoretical grounds (Webb,
1966) for the interaction between a spherically stratified
absorbing atmosphere and incident solar radiation. On the
other hand, the data presented in Figure 4.23 indicate the
presence of dynamic-type variations, even in these mean
data, which effectively eliminate the concept of total solar
radiation control of the stratospheric circulation.

In addition to the data summarized in the above para-
graphs, the cryogenic sampler has very recently produced
data indicating moisture concentrations in the ten parts per
million range near the stratopause and a very surprising
several hundred parts per million in carbon dioxide con-

centration. Such minor constituents offer an alternate physical process through which the observed highly variable character of the stratospheric circulation may be established and thus are of critical interest to the atmospheric scientists.

4.4 Thermospheric Circulation

Systematic motions of the neutral atmosphere in the region above approximately 80 km altitude are here termed the *thermospheric circulation*. This atmospheric region has thus far not been explored from the synoptic meteorological point of view, so that our knowledge of the meteorology of that region is very meager. This state of affairs is in large part the result of an inordinate interest in the impact of free electrons on our communications technology and in part the result of an incomprehensible attitude on the part of many meteorologists that the upper atmosphere is of no consequence in world affairs. It is time for a more realistic approach in the atmospheric sciences; namely, one in which the atmosphere is viewed as a unified whole.

Attempts to obtain another segment of the required data have already been initiated. The *meteor trail radar technique* of wind measurement appears to be a desirable method for synoptic inspection of the base of the thermospheric circulation in the 80-110 km altitude region. At a summer institute held by the Physics Department of the University of Texas at El Paso in August 1970, leading meteor trail radar scientists were assembled from over the world to discuss the observational technique and its use in a global synoptic observational system. The proceedings of that summer institute are being published (Webb, 1971) to provide easily available information on all aspects of the observational technique for use by scientists who may cooperate in implementation of a meteor trail radar network.

One of the most interesting results which have been produced by the several research inspections of the lower thermosphere which have been accomplished with the meteor trail radar technique is the indication that there exists a *systematic meridional flow* from summer to winter polar regions. As is indicated by the data presented in Figure 4.24, the amplitude of this circulation system is approximately 10 m/sec^{-1}. Such a result was not altogether unexpected, since thermodynamic reasoning had already inferred *upward motions* in summer high latitudes at mesopause altitudes ($\sim$80 km) to produce the lowest temperatures in the atmosphere and *downward motions* across the mesopause in winter high latitudes to produce the rather high temperatures observed in this location which is devoid of solar heating. These initial data would then indicate that the White Sands Missile Range meteor trail radar system will observe such an annual variation in meridional circulation structure.

The second interesting fact which has been observed

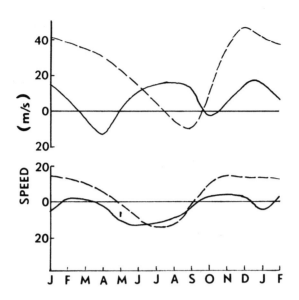

Figure 4.24. Monthly mean zonal (upper curves) and meridional (lower curves) winds at 92 km altitude over Jodrell Bank (53°N, 2°W) and at 90 km altitude over Adelaide (35°S, 138°E) are indicated by the solid and dashed curves respectively.

in initial exploration of the lower thermosphere is the fact that at upper middle latitudes (Jodrell Bank) there appears to exist a *semiannual variation* in the zonal wind field. These data indicate westerly winds at solstice times and easterly winds at equinox times. This is interesting since an inverse phase semiannual wind variation in equatorial regions has already been observed at the stratopause level (~50 km). This latter circulation system is known to be associated with the winter storm periods, so it is natural to inquire as to whether these semiannual variations of the 50 and 100 km levels might simply be components of a comprehensive circulation system.

Data from the meteor trail radar system currently under construction at White Sands Missile Range will then provide much needed new information on the global circulation system of the base of the thermosphere. Parallel with the activation of this new observational system will be an effort to implement synoptic-type observations at all existing meteor trail radar stations and to foster development of additional stations. These data will be required before complete understanding of interactions between the neutral and electrical atmospheres discussed in Chapter 6 can be achieved and are also fundamental to the propagation characteristics of waves discussed in Chapter 7 and to the overall structure of the atmospheric system.

DIFFUSION STRUCTURE

If the atmosphere were *quiet*, the gravitational field would *separate* the various atmospheric gases according to their molecular weights. The heaviest gases would exhibit the highest concentrations at the surface and the relative concentrations would vary with height. This situation does not exist in the atmosphere in the lower 100 km, but rather the gases are observed to be rather *well mixed*. Since the gases which make up the atmosphere exhibit a wide range of molecular weights, these data are assumed to demonstrate that in general eddy diffusion dominates small-scale transport processes in the lower 100 km of the atmosphere.

Under static conditions, the vertical structure of an atmosphere in a gravitational field is given by the *hydrostatic relation*

$$dp = -\rho \, g \, dh$$

where p is pressure, ρ is density, g is gravity and h is height. In an isothermal atmosphere this relation integrates to yield

$$p = p_0 \, e^{-h/H},$$

where p_0 is the pressure at the base of a layer and the scale height is $H = RT/mg$, R is the gas constant, T is the absolute temperature of the layer and m is the molecular weight. This condition would prevail if *molecular diffusion* should dominate the atmospheric transport processes, and each of the atmospheric gases would assume its characteristic static profile.

Molecular diffusion transport in the atmosphere over El Paso is specified rather accurately by the coefficient (K_m) structure illustrated by the solid curve of Figure 5.1 which is derived from other atmospheric variables (Webb, 1970a). The dashed curve of Figure 5.1 illustrates the mean eddy transport coefficient (K_e) which should be roughly characteristic of a nice winter day over El Paso. This transport coefficient is composed of a complete spectrum of eddy sizes which are hypothesized to have an energy spectrum (q) structure illustrated by the relation

$$q \propto \varepsilon^{2/3} k^{-5/3},$$

where ε is the viscous dissipation rate per unit mass of the fluid and k is the *wave number*. Eddy transport is far more variable than molecular transport, with order of magnitude and larger variations to be expected above and below the surface intensities indicated in Figure 5.1.

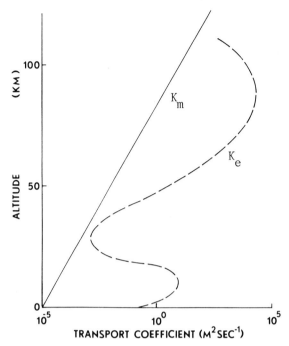

Figure 5.1. Model profiles of atmospheric molecular (solid curve) and eddy (dashed curve) transport coefficients based on Lettau (1951) in the neutral atmosphere and Booker (1956) in the ionosphere.

These two transport modes produce vertical fluxes of heat after the fashion indicated by

$$F_m = - K_m \rho c_p \left(\frac{\partial T}{\partial z}\right)$$

and

$$F_e = - K_e \rho c_p (\gamma - \Gamma) ,$$

where c_p is the specific heat at constant pressure, γ is the ambient lapse rate (rate of temperature decrease with height) and Γ is the adiabatic lapse rate (-1°C/100 m). It is clear from the data presented in Figure 5.1. that eddy transport will dominate the vertical diffusion heat transport situation, tending to force the ambient lapse rate to become more nearly adiabatic where the eddy mixing is strong.

Eddy mixing serves to transport any *minor constituent* if its local concentration lapse rate is other than the total atmospheric pressure lapse rate. This is stated as

$$F_q = - k_e \frac{\partial q}{\partial z}$$

where q is the density of the minor constituent. This simple

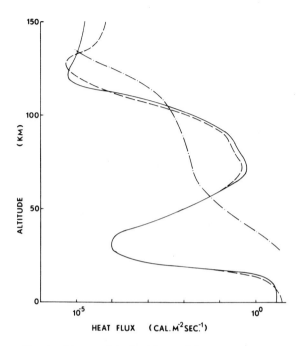

Figure 5.2. Heat flux which the eddy transport coefficients of Figure 5.1. indicate will flow downward for summer (dashed curve) and winter (solid curve) cases. The dash-dot curve is an estimate of the solar radiant energy absorbed above each level, based on the work of Hinteregger, Hall and Schmidtke (1965), Johnson (1966) and Lettau (1951).

picture is complicated in the atmosphere by the fact that K_e is variable in space and time and that there are numerous sources and sinks for most atmospheric constituents. A good example is *water vapor*, where the principal source is at the earth's surface, with major sinks in the upper atmosphere (60-100 km) through photodissociation processes and in the troposphere through the precipitation process. Even these complications fail to delineate the observed water vapor structure, so we can be sure that there are additional sources and sinks for this substance. Another good example is the observed *atmospheric ozone* structure, where the principal source is in the upper atmosphere (65-90 km) and the reduced eddy transport of the lower stratosphere re- sults in maximum accumulation before the ozone is transported downward by strong eddy diffusion in the troposphere to be burned at the ground.

Based on the model transport coefficients of Figure 5.1. a typical estimate of atmospheric heat transport applicable for a summer noontime atmosphere over El Paso is presented in Figure 5.2. This transport curve is then differentiated to obtain the heat input obtained for the atmosphere by

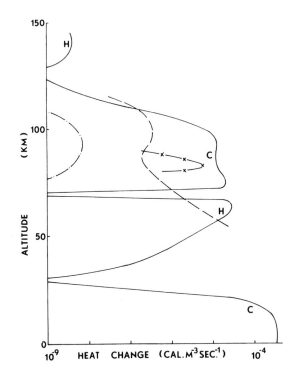

Figure 5.3. Divergence (C) and convergence (H) of vertical
eddy transported heat (solid curve), along with the solar
radiant (dashed), chemical (dash-cross) and electrical
(dash-dot) heat inputs.

this mechanism as is illustrated in Figure 5.3. These
values indicate the importance of eddy transport processes
in establishing the thermal structure of the atmosphere.
The hot lower thermosphere and stratopause regions clearly
owe part of their heat to local accumulations which result
from variable vertical eddy transport rates. In addition,
theoretical studies of the impact of eddy diffusion on
minor constituent structure and chemical reaction rates have
illustrated the importance of abandoning archaic static
atmosphere concepts for a more dynamic situation.

ELECTRICAL STRUCTURE

The earth has a *complex* electrical structure. El Paso's
low-latitude location provides for a representative sample
of the electrical phenomena which are characteristic of
that region. There is a fair sample of the obvious lightning, and with suitable sensors the electrical structure of
the El Paso area is generally found to be of great scientific interest. A most important new concept in this field
concerns a unity which now appears to be a dominant feature
of the earth's electrical structure (Webb, 1968a, 1968b,
1969b, 1970b).

The atmosphere is a semiconducting medium which is
rendered electrically conducting through the presence of
free electrons and ions. These charged particles produce
an electrical conductivity structure over El Paso at summer
noontime of the type illustrated in Figure 6.1. The facility with which charged particles in a gas transport electricity is determined by the temperature and collision frequency structures of the gas. It is clear that the atmosphere will exhibit a wide range of electrical characteristics, ranging from a good conductor in the upper portions
to a rather poor conductor in the troposphere.

The lower atmosphere electrical structure (Chalmers,
1967) in fair weather is characterized by high impedances
and voltages. With resistance of almost 10^{13} ohm meters
and surface vertical potential gradients of 50-150 volts
per meter, the troposphere generally obeys Ohm's Law
(E=IR) and delivers a downward-directed current density of
approximately 10^{-12} amperes per square meter. Such a
current involves the deposit of roughly a million positively
charged particles per square meter per second. It is important to note that, when summed over the enitre fair-
weather area of the earth, this current density constitutes
a total vertical current of approximately 1500 amperes.

A basic reason for this rather steady state of affairs
is the simple fact that the earth in its fair weather areas
is continually in possession (for the more than 200 years
of record) of an approximately 400,000 Coulombs of negative
charge. That is, the current flow onto the earth in fair
weather appears to be produced by attraction of positive
ions in the atmosphere to the negative charges on the
earth's surface. El Paso has its fair share of about 10^{-10}
Coulombs (10^8 electrons) per square meter of negative surface charge in fair weather.

Clearly, a total *resident charge* in the 10^5 Coulomb
range will be very quickly eliminated by a 10^3 ampere
(1 ampere=1 Coulomb per second) discharge current (namely;
in the order of 100 seconds) if there were no return current
path and no electromotive force to keep the current flowing.

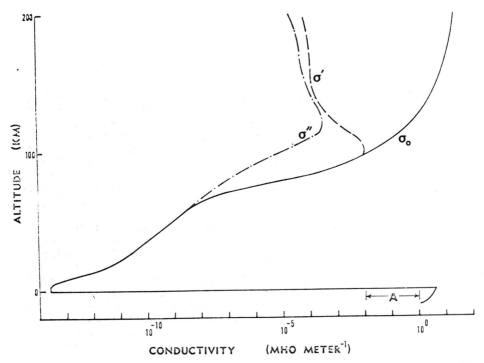

Figure 6.1. Typical noontime conductivity structure over El Paso for the specific (σ_0), Pederson (σ') and Hall (σ'') components after Cole and Pierce (1965) from the surface of the ocean to 100 km and Hanson (1965) above 100 km. The symbol A refers to the range of variable conductivity of the earth's crust.

Thunderstorms, with their very special lightning discharges, have been identified as the site of the return current path and have been postulated to be the motivating force which maintains the tropospheric electrical structure.

The simple *spherical condensor model* of global tropospheric electrification which satisfied the above data was sufficient for the pre-space era. Above approximately 50 km, however, the electrical situation is far more complex. The noontime vertical profile of electrical conductivity over El Paso is modeled in Figure 6.1., based on the relation (Hanson, 1965)

$$\sigma_0 = \frac{n\ q^2}{m_e\ \gamma_e} + \frac{n\ q^2}{m_i\ \gamma_i}$$

for the *specific conductivity* (along the magnetic field), where n is the charged particle number density, q is the individual particle electrical charge, γ is the collision frequency and m_e and m_i are the electron and ion masses, respectively. The *Pederson conductivity* (normal to the

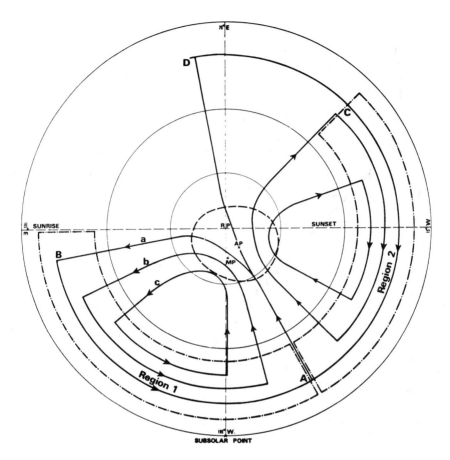

Figure 6.2. Equatorial plane projection of the dynamo current system. Regions 1 and 2 (enclosed by dash-dot curves) represent regions of electromotive force generated by vertical tidal motions. Positions of the rotational (RP), magnetic (MP) and auroral (AP) poles for the Northern Hemisphere are indicated. The dashed circular curve represents the center line of the auroral zone.

magnetic field and along any electric field) is determined by

$$\sigma' = \frac{n\,q^2\,\gamma_e}{m_e\,(\gamma_e^2 - \omega_e^2)}\quad \frac{n\,q^2\,\gamma_i}{m_i\,(\gamma_i^2 - \omega_i^2)}\;,$$

where $\omega = qB/m$ is the gyrofrequency of the charged particle around the magnetic field line and B is the intensity of the magnetic field. The *Hall conductivity* (normal to both magnetic and electric field vectors) is determined by the relation

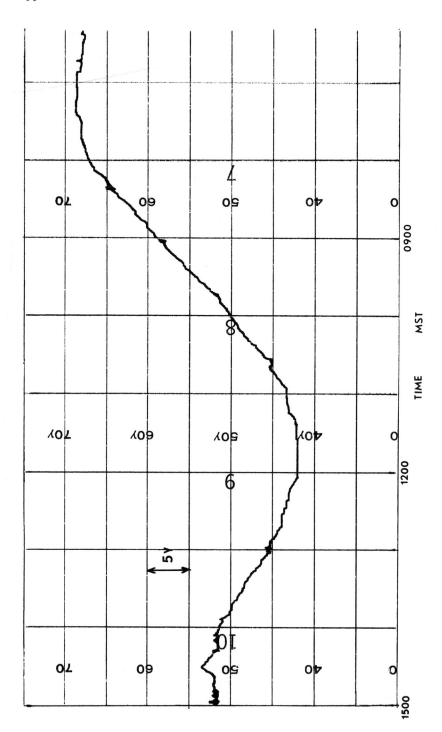

Figure 6.3. Diurnal variations in surface magnetic field intensity observed at White Sands Missile Range on 16 May 1970.

$$\sigma'' = - \frac{n\,q^2\,\omega_e}{m_e\,(\gamma_e^2 - \omega_e^2)} + \frac{n\,q^2\,\omega_i}{m_i\,(\gamma_i^2 - \omega_i^2)}$$

Positive and negative ions are the principal contributors to this conductivity structure in the atmosphere below 50 km, while above 70 km electrons carry most of the current.

As a result of *systematic vertical motions* of the stratopause tidal circulations discussed in Section 4.3, differential transport of positive ions and electrons across the earth's magnetic field at low latitudes produces through the Hall effect (Fejer, 1965) a horizontal current system in the 100 km region of approximately 10^{-5} amperes per square meter density flowing from the west during the morning hours in both hemispheres as is illustrated in Figure 6.2. These *dynamo currents* (Chapman and Bartels, 1940) reverse direction and are less intense by an order of magnitude or more during the evening and at night. The total current flowing in each of the two sunlit dynamo circuits falls in the 100,000-500,000 ampere range, depending principally on the season of the year.

The presence of the dynamo current systems over El Paso is best evidenced by the diurnal variations which they induce in the earth's magnetic field. The earth's permanent magnetic field in the El Paso area has a declination of 11° east, a dip angle of 58° and an intensity of .5 gauss. A typical rubidium vapor magnetometer record of the total field intensity is illustrated in Figure 6.3., showing a characteristic diurnal variation of approximately 50 gammas (1 gamma = 10^{-5} gauss), with maximum reduction of the earth's magnetic field intensity when the overhead dynamo current system is a maximum at approximately 11 a.m.

Electric currents also flow upward and downward along the magnetic field above the dynamo currents. These currents are in response to electric potential differences between hemispheres which are established by the dynamo currents. Energetic particles precipitate from the magnetosphere and interact with upper atmospheric particles to ionize and excite and thus contribute to the air glow emission which produces a weak visible glow in the dynamo current region.

WAVE STRUCTURE

The generation, propagation and dissipation of *waves* in the atmospheric medium represents one of the more sophisticated means of storing and transporting energy. In the lower atmosphere the amount of energy involved in wave motions is generally quite small when compared with the total energy density, but this situation changes rapidly with height so that in the mesosphere and lower thermosphere this mode of energy processing may dominate the dynamics of the region. The total impact of waves on atmospheric structure is as yet uncertain, but clearly their importance is sufficient to require a general understanding of atmospheric response to these phenomena.

Most familiar of atmospheric waves are *sound waves*, which have the general characteristic that they propagate with the speed of sound when their pressure perturbations are small when compared with the ambient pressure. The speed of sound in a gaseous medium is given generally by the relation

$$v = (\gamma \; RT)^{1/2}$$

where γ is the ratio of specific heat at constant pressure to that at constant volume, R is the gas constant characteristic of the gas and T is the absolute temperature. For dry air this relation reduces to

$$v = 20.06 \; \sqrt{T} \quad (m/s)$$

The motions predicted by these relations are those which will occur relative to a coordinate system which is moving with the wind or in a fixed coordinate system in still air.

Typical examples of these *Eulerian* speed-of-sound profiles as observed over the El Paso area are illustrated in Figure 7.1 (Webb and Jenkins, 1962). A most obvious feature of these profiles is the general *waveguide* nature of the speed of sound structure of the atmosphere. Sound waves generated in or entering these regions with the proper angles with the horizontal (<30°) will be ducted and will travel over great distances without reflection at the earth's surface or attenuation in the thin upper atmosphere. Divergence of the sound wave will be circular under the assumptions outlined above.

Sound waves *lose energy* as they propagate through the atmosphere. This energy loss is much greater for higher frequencies and at higher altitudes. Dean (1959) has calculated the losses according to the following relation

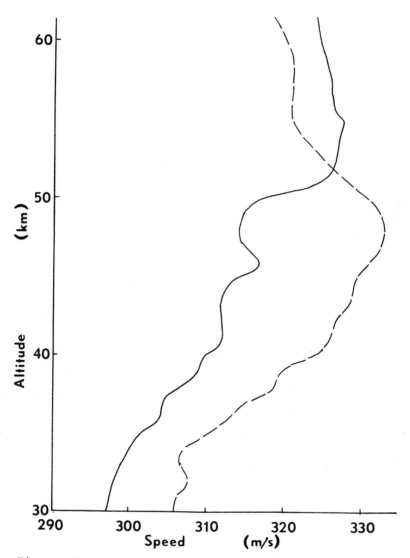

Figure 7.1. Static speed-of-sound profiles observed over White Sands Missile Range on 10 December 1968 (solid curve) and on 19 June 1968 (dashed curve).

$$\alpha = \frac{2.5 \times 10^{-7}(T^*)^{0.2}f^2}{p^*} + \frac{5.3 \times 10^{-4}(T^*)^3 f}{p^*}$$

$$+ \frac{10^2 f}{(T^*)^{2.5} e^{3.18/T^*}} \times \frac{2f_o f}{f^2 + f_o^2} \quad (\frac{db}{mile}) \quad ,$$

where f is the frequency in cycles per second and f_o is
given by

$$f_o = \frac{P*}{(T*)^{0.8}} \ [\ 900 \ (\frac{wT*}{P*}) \ + \ 500 \ \frac{wT*}{P*} \ + \ 50 \ ],$$

$T* = \theta/273$, T is in degrees Kelvin, P* is pressure in atmospheres and w is absolute humidity in grams per cubic meters.
Water vapor operates to absorb sound waves, with an increase
in sound absorption at 1 Hz from 6.7×10^{-5} decibels per mile
at 100% relative humidity at the surface to 8.6×10^{-5}
decibels per mile at 10% relative humidity at the surface.
At 100 Hz these values are 1.2×10^{-2} and 2.0×10^{-1} respectively.
    Increase in absorption with altitude is even more pro-
nounced, with the 1 Hz case increasing from the surface 100%
value to 4.8×10^{-2} at 45 km at a relative humidity of 10%.
The 100 Hz absorption increases to 4.7×10^{0} over the same
region. Clearly, high frequencies are limited in range,
particularly in the vertical direction. Intense sounds such
as those from explosions, supersonic flight, and turbulent
wind fluctuations fall into the 1 Hz and less frequency
range and thus may be detected at great distances from their
sources. In fact, the atmosphere contains a rather large
amount of such energy, with observed background sound pressure levels at the 20 km level in the .1-10 dynes per cm^2
range (Webb, 1959). These high sound levels are inaudible
to the ear, and their origin and propagation characteristics
are not now well known. While their importance in establishing the structure of surface layers is relatively minor,
their propagation into the upper atmosphere would be an
altogether different matter, with the wave energy component
possibly dominating the total energy content of the ambient
air.
    The speed-of-sound structure of the atmosphere is con-
siderably *complicated* by the fact that the wind is a vector
quantity (**U**). The sound velocity profile (**V**) relative to a
coordinate system fixed to the earth's surface is then
given by the relation

$$\mathbf{V} = v + \mathbf{U}$$

A principal result of this relationship is that the sound
wave front does not always travel normal to itself, and
thus the usual ray-tracing techniques of geometric optics
lead to erroneous results. In practice, it is common to
resolve these effects into translational and rotational
components which will produce a relatively good approximation of the actual ray path.
    Characteristic vertical speed-of-sound profiles for
the El Paso area in *summer and winter* for sound waves
traveling vertically, from the east, and from the west
are presented in Figure 7.2. In winter the strong westerly
winds strengthen the lower duct for sound waves traveling

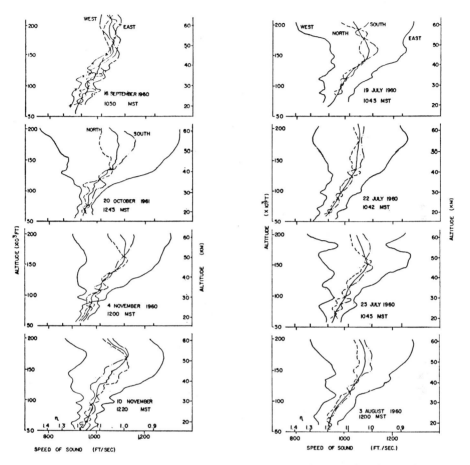

Figure 7.2. Typical speed-of-sound profiles during winter and summer over White Sands Missile Range from Webb and Jenkins (1962).

from the west, while virtually eliminating the duct for sound waves traveling from the east. In summer the wind effect is roughly the opposite, altough the impact on the lower duct is not so great and that on the upper duct is greater. In all cases these profiles exhibit a considerable amount of detail structure (Webb, 1962, 1965).

Another large class of atmospheric waves are *gravity waves*. These waves obtain their names from the fact that their motions are controlled principally by the energy of the gravity field of the earth. They follow a different set of rules relative to generation, propagation, and dissipation. Obvious examples of these waves include *tidal waves* resulting from diurnal heat inputs and gravitational effects, *lee waves* caused by wind flow across topographic obstructions, and pressure perturbations radiated away from intensifying

and decaying atmospheric pressure systems. These gravity
waves are low in frequency, with an upper limit set by the
Brunt-Väisälä natural frequency of oscillation of the atmosphere which is given by the relation

$$\nu = \frac{g}{\theta} \frac{\partial \theta}{\partial z} ,$$

where θ is the potential temperature defined by

$$\theta = T(1000/p_o)^{R/c_p}.$$

There is a final class of waves which are evanescent
(*imaginary*) in which there is no propagation of wave energy
out of the region in which it is deposited.
   Lee waves are a special type of gravity wave. They
frequent the El Paso area during winter and spring as winter
westerly winds are forced to move vertically to cross the
Franklin Mountains. These waves were discussed in Section
4.2. Of special import is the fact that these waves exhibit *large accelerations*, and with such accelerations it is
to be expected that internal propagating waves will be
generated in the fluid. Such waves will transport energy
into other regions of the atmosphere, and in some cases the
dissipation of this wave energy may have significant impact
on the structure of the atmosphere (Jones, 1969).
   There appear to be many other sources of propagating
waves in the atmosphere, and the significance of this mode
of energy transport is just beginning to be fully realized.
Particularly in the case of upper atmospheric structure is
it likely that wave phenomena will be of importance in
determining atmospheric structure.

DESIGN OF A MESOMETEOROLOGICAL NETWORK FOR EL PASO

The nature of the data acquisition and processing problems which are associated with meteorological data required
for ecological purposes in our modern economy demands that
the entire process be automated. The mesonet model outlined
here incorporates use of a digital computer for collection,
evaluation, storage and presentation of the data and, where
possible, involves automation of the sensors and telemetry
systems to reduce manual operation of the system to a minimum.

Acquisition of more sophisticated sets of meteorological
data for application to ecological problems would be of little
value unless those data are efficiently applied toward solution of those ecological problems. It is essential that
parallel with the development of such a data system an educational system be implemented which will assure an aware-
ness by our technological managers of the virtue and useful-
ness of the data. For these reasons it is proposed that the
expanded mesometeorological system outlined here include a
data acquisition and processing system which is directly associated with the public school systems of the El Paso metropolitan area. Wherever possible the system would be located
at and operated by the science departments of these institu-
tions, with the resulting data presentations provided to those
law enforcement, medical and control organizations as well
as others in our technology who have need for environmental
data.

To facilitate development of such a unified system of
data storage, processing and calculation for ecological purposes, a unique grid system for mesometeorological work is
presented for the El Paso metropolitan area. This grid is
illustrated in Figure 8.1, along with an outline of the city
and the location of the principal observational locations
which would make up the initial network. Kilometer units
are used in the coordinates of this grid system.

The network design specifies that each of the public
high schools plus certain other locations in the metropolitan area have standardized observational stations located on
their grounds and that these basic network stations be tied
to a central point by telephone data links. Those parameters
observed will initially be meteorological measurements of
wind, temperature, precipitation, thermal stability and wind
shear, with provision for later incorporation of other environmental measurements as required. Since the sampling
rates for these data needs to be highly variable, the data
transmission and processing systems should be able to provide
for routine (say at hourly intervals) pickup of the several
parameters being measured with a capacity of increased sampling rates (say at one-minute intervals) during periods of

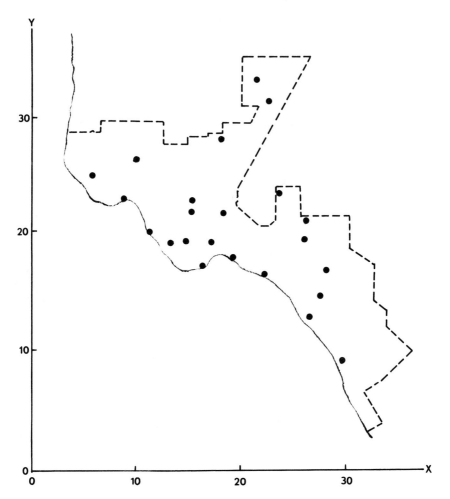

Figure 8.1.   The El Paso mesometeorological grid.   Coordinates
are in kilometers.

---

special import.
     The network design specifies that the central point for
collection, storage, processing and presentation of the data
from this mesomet network be the Computation Center at the
University of Texas at El Paso.   The Control Data Corporation
3100 computer in operation there can accept data from the
mesonet and special observational stations and process these
raw measurements into forms which are optimum for application
to particular ecological problems.   An example of the initial
mode of operation of the entire system is presented in this
Chapter, using the design locations and altitudes of the
mesomet network stations presented in Table 8.1 to derive
the altitudes of the grid points and to present a topographic

map of the area derived from these data. Finally, an air
pollution problem is evaluated in the usual Pasquill-Gifford
mode to delineate the nature of the additional computations
which will be required when more sophisticated mesometeorological data become available for this application. These
examples are designed to illustrate the limitations of current methods of handling the air pollution problem and indicate the improvement which may be expected with implementation of the more comprehensive meteorological observational
system proposed here.

The techniques presented here for processing and applying atmospheric data to El Paso's ecological problems must
be regarded as an extreme simplification of the programs
which will finally be required. The approach does represent a firm step, however, and thus lends itself to improvements which can best be derived through experience.

8.1 Measurements

Basic measurements in the mesomet network will be
meteorological in nature, with provision for additional observations for minor constituents, visibility, etc., as
needed. Meteorological data currently required for local
atmospheric control purposes include measurements of at
least precipitation, temperature, humidity, wind and boundary
layer thermal and wind gradients. The measuring systems must
be tailored to the telemetry system, and there is at present
no standard technique available. The following general approach is indicated:
a. Precipitation: Tipping bucket raingages provide a standard technique for precipitation measurement which should
prove generally adequate for the El Paso mesonet. Rainfall
is ducted from the collection opening into a weighting device which activates an electrical circuit each time it dumps.
This dumping event is usually calibrated to occur when the
load is .01 inch of water in the collection opening. In the
mesomet case this calibration should be changed to an event
for each .01 cm of water in the collection opening. The
number of times the measuring device tips is the data to be
telemetered.
b. Temperature and humidity: Dry and wet thermometer read-
ings are usually obtained under aspirated conditions through
use of similar semiconductor sensors. Remote sensing of
these parameters are currently accomplished by standard de-
vices with known characteristics in many applications, and
these systems should be used here if the telemetry systems
can be matched.
c. Wind: Anemometer systems are available in standard form.
Instruments with sensitivities of the Beckman-Whitley level
or better should be used.
d. Thermal and Wind Gradients: Measurement of vertical
gradients in the temperature and wind fields provides some
significant new problems. It is proposed that for the El
Paso mesomet network these measurements be made in the 2-20

Table 8.1. Mesometeorological network stations for El Paso.
Locations are in kilometers on the mesonet grid and elevations
above MSL are in meters.

Station	X	Y	Z
Andress	21.7	34.0	1204
Ascarate	22.6	16.4	1122
Austin	18.5	21.6	1200
Bel Air	28.2	15.8	1219
Bowie	16.4	16.7	1131
Burges	24.8	20.2	1198
Club	5.1	26.2	1143
Coliseum	19.2	17.9	1128
Coronado	9.4	25.8	1225
Dam	11.2	20.1	1138
Eastwood	27.5	18.7	1207
El Paso	14.6	18.8	1226
Irvin	21.9	30.7	1189
KROD	15.6	21.2	1529
KTSM	15.4	22.4	1713
Park	18.5	28.2	1244
Sunland	8.1	22.4	1136
Tech	16.5	19.7	1214
Thomas	25.6	14.4	1115
UTEP	13.4	18.4	1165
Water	20.6	24.8	1180
Weather	24.1	21.6	1195
Ysleta	28.7	10.7	1122

meter (6-60 feet) height range, using thermocouple and hot
wire systems for measurement of temperature and wind speed
differences, respectively. The latter approximation to the
true wind gradient is considered justified in view of the
experimental difficulties of a more sophisticated approach
and the chaotic turbulence structure which characterizes
the boundary layer.

The basic El Paso mesomet network data will then con-
sist of five items of data which must be processed for telemetry and insertion into the computer. Additional provision
should be made for other measurements which will surely be
optimumly made at the mesomet sites.

8.2   Data Processing

Raw meteorological and/or air pollution data will be
provided to the computer from the mesomet stations illustrated in Figure 8.1 and listed in Table 8.1. In view of
the many uses to which these data might be applied, these
inhomogeneous data should be interpolated and extrapolated
onto the standard mesomet grid, and this form of the data
should be used for preparation of presentations or for
additional ecological calculations. This procedure will

have the virtue of providing the best meteorological data
available in a unified format ready for computer application
to a variety of environmental problems.

The first step, then, is to process the raw mesomet
data onto the mesonet grid. This is accomplished after the
mode developed by Inman (1970) for the National Severe
Storms Forecast Center's mesometeorological network. The
procedure involves first estimates of the grid point values
of the parametric field being analysed. In general, the
overall mean of the observational data provides a satisfactory initial estimate. The field value (z_k) at an observational point is then interpolated from the relation

$$z_k = z_1 + (z_4 - z_1) \frac{\Delta y}{b} + (z_2 - z_1) \frac{\Delta x}{a}$$

$$- (z_2 - z_3 + z_4 - z_1) \frac{\Delta x}{a} \frac{\Delta y}{b} ,$$

where the grid points of the segment in which p is located
are numbered consecutively counterclockwise from the point
from which Δx and Δy are measured and a and b are the x and
y grid point spacings (1 km) respectively.

The difference (D_k) between this computed value and the
observed value is then used to derive grid point corrections
($C_{i,j}$) based on the relation

$$C_{i,j} = \frac{\sum_{k=1}^{n} W_k D_k}{\sum_{k=1}^{n} W_k} ,$$

where k is the number of observations involved. W is a
distance-dependent weighting factor defined by

$$w = \frac{R^2 - d^2}{R^2 + d^2} ,$$

where d is the distance between the grid point and the observation point and R is the radius of influence. In our
case R is set equal to 10 km.

A new estimate of the parametric value at each nearest
grid point is then obtained from the relation

$$z' = z + C'.$$

This process is iterated until C' becomes negligible and the
parametric field is considered to be a best estimate.

Derivation of parametric values for individual grid
points includes information contained in the data from each
observational station within 10 km, with emphasis on those
stations which are located closest to the grid point. Clearly, any such process of surface fitting to the data will

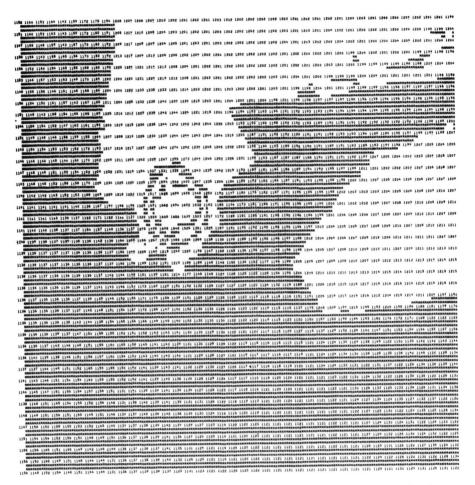

Figure 8.2 Topographic map at 100-meter intervals of the
El Paso area derived from the station altitude data of Table
8.1 processed with smoothing through the program outlined
in Appendix I. Altitudes in meters MSL are printed at km
grid points.

tend to eliminate small-scale features contained in the
original atmospheric structure. Since the severe limitations which are always inherent in meteorological observational systems will already have filtered much of this detail structure out of the information available to the computer, it is desirable to minimize any further filtering
effects.

The Fortran program used to accomplish this processing
of observational data is presented in Appendix I. This
program is available at the University of Texas at El Paso
Computer Center for conversion of any inhomogeneous set of
data onto any uniform grid.

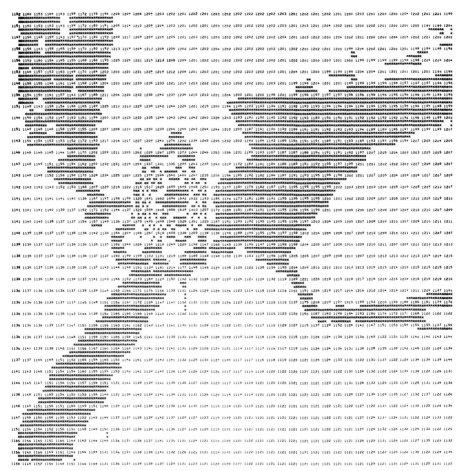

Figure 8.3 Topographic map at 50-meter intervals of the
El Paso area derived from the station altitude data of
Appendix I. Altitudes in meters MSL are printed at km
grid points.

As an example of the technique, the altitude data for
the mesonet stations listed in Table 8.1 were processed
through the program. The computer output for each 1 km grid
point of these input station altitude data was then used to
derive the contour maps presented in Figures 8.2, 8.3, 8.4
and 8.5. Comparison with the crude topographic map of
Figure 2.1 or with more detailed topographic maps illustrates
the degree to which this total process of data handling produces a surface which represents the topography of the El
Paso area. While these results are clearly disturbing, it
must be remembered that the step which we are attempting
to take is to advance from the current practice of employ-
ing one value of a required parameter to represent the en-

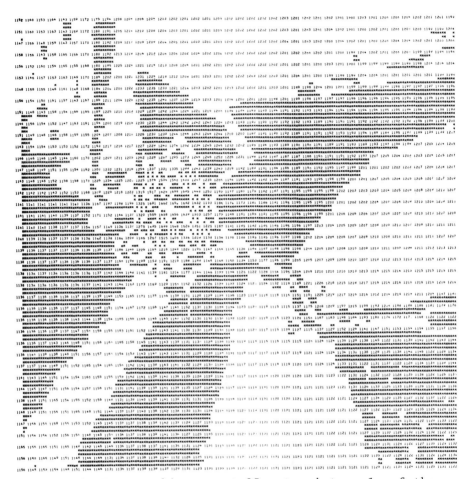

Figure 8.4 Topographic map at 25-meter intervals of the El Paso area derived from the station altitude data of Table 8.1 processed with smoothing through the program outlined in Appendix I. Altitudes in meters MSL are printed at km grid points.

tire El Paso area to a more sophisticated use of a complex environmental surface.

Since the atmosphere is known to exhibit occasional complexities of the same order as that illustrated in Figure 2.1, it is obvious that certain small scale aspects of atmospheric structure cannot be studied through data obtained with the mesomet network outlined here. Improvements in resolution will generally mean more observational points with their attendant economic problems. It seems probable at this point that the data processing techniques presented here will prove quite satisfactory for the observational network outlined in Figure 8.1 and Table 8.1.

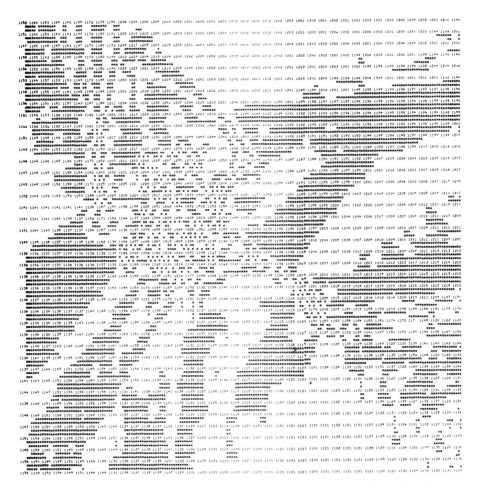

Figure 8.5 Topographic map at 10-meter intervals of the
El Paso area derived from the station altitude data of
Table 8.1 processed with smoothing through the program
outlined in Appendix I. Altitudes in meters MSL are printed
at km grid points.

8.3 Air Pollution Dispersion

Polluting material released into the atmosphere is, in
most cases, dispersed through and removed from the atmosphere
through a variety of physical processes. The most obvious
of these cleansing mechanisms is the precipitation process,
which traps gases and particulates into droplets which are
separated from the atmosphere to the ground by the gravity
field. Coagulation and sedimentation processes in the ab-
sence of liquid water are also effective cleansing mechanisms.
Most polluting events which are of significant interest

involve rather high concentrations at the source, and before the atmospheric cleansing mechanisms can be really
effective in eliminating the material it will be spread
through a relatively large volume of the atmosphere. This
dilution is really a cleansing mechanism from the polluter's
point of view, since there is in each case a threshold concentration for each material below which the presence of
that minor constituent is not objectionable to man. The
fact that polluting material does disperse makes it possible
for the atmosphere to be used as a dumping ground for human
wastes. Otherwise, the polluter would soon be enmeshed in
his own effluent.

The dispersing mechanisms active in the atmosphere are
of prime interest to the atmospheric scientist for a number of reasons. The basic dispersion mechanism in any gas
is molecular diffusion, through which individual contaminating molecules random-walk away from the source as a result of the Brownian motions of the gas. While always a
potent mixing mechanism, molecular diffusion is generally
overpowered by other transport mechanisms in the lower atmosphere.

The most powerful dispersion mechanism in the atmosphere
is the general circulation. The wind dilutes a given source
output by introducing that particular amount of material
into a larger volume of the atmosphere. Strong winds then
spread the pollution around very effectively, and there are
few problems with local pollution during the windy seasons.

A dispersion mechanism of intermediate importance is
dispersion through eddy mixing. Turbulent eddies in the
atmospheric flow field effectively transport minor constituents in concentration gradients with time constants which
are much smaller than the corresponding molecular time
constants. This process has been discussed in detail in
Chapter 5, and integration of the basic diffusion equation
provides the basic Fickian technique of approximating the
transport provided by this physical process.

The use of tall smokestacks along with eddy transport
has been a favored approach of industry in spreading their
pollution around. This technique has received much study,
and the most commonly employed method for handling the
problem is illustrated here. It is important to note that a
smooth earth and a stratified atmosphere which is homogeneous
in lateral directions are generally assumed. While this is
a reasonable first approximation, it is certain that this
approximation to true conditions will fail under some conditions, and it is highly probable that in the El Paso area
a more sophisticated picture of atmospheric structure will
be required before results adequate for an advanced technology are achieved.

The relationship usually used for calculating the dimensions of the plume generated by a continuously emitting
stack of source strength Q at a height h which produces a
plume axis at height H after thermodynamic equilibrium is
achieved is expressed as (Pasquill; 1962; Slade; 1968)

$$q = \frac{Q}{2\pi\sigma_y\sigma_z\overline{U}} e^{-y^2/2\sigma_y^2} e^{-(z-h)^2/2\sigma_z^2} e^{-(z+h)^2/2\sigma_z^2} ,$$

where σ_y and σ_z are the rms deviations of displacements normal to the plume axis with y horizontal and H measured vertically along z.

The information desired in air pollution work is generally the concentration at the surface (q'), in which case the above equation simplifies to

$$q' = \frac{Q}{2\pi\sigma_y\sigma_z\overline{U}} e^{-y^2/2\sigma_y^2} e^{-H^2/2\sigma_z^2} ,$$

A solution of this equation has been programed at the University of Texas at El Paso Computation Center after the method employed by Heimbach (1970) at the Meteorology Department of the University of Oklahoma. The program is presented in Appendix II.

For illustrative purposes calculations were carried out for a special case. A source strength (Q) of 1000 grams per second placed at an effective height (H) of 229 meters dispersed into a 5 mps mean wind ($\overline{U}$) was employed under various atmospheric stability conditions (A-Extremely unstable conditions; B-Moderately unstable conditions; C-Slightly unstable conditions; D-Neutral conditions; E-Slightly stable conditions; F- Moderately stable conditions) and various grid sizes to illustrate the factors involved. The results of these calculations are presented in Figures 8.6-8.24. These computer printouts are coded as follows:

$$q' = - : \quad < 10^{-13} \qquad (g\ m^{-3})$$

$$1 : \quad 10^{-11} - 10^{-13} \qquad ''$$

$$2 : \quad 10^{-10} - 10^{-11} \qquad ''$$

$$3 : \quad 10^{-9} - 10^{-10} \qquad ''$$

$$4 : \quad 10^{-8} - 10^{-9} \qquad ''$$

$$5 : \quad 10^{-7} - 10^{-8} \qquad ''$$

$$6 : \quad 10^{-6} - 10^{-7} \qquad ''$$

$$7 : \quad 10^{-5} - 10^{-6} \qquad ''$$

$$8 : \quad 10^{-4} - 10^{-5} \qquad ''$$

$$9 : \quad 10^{-3} - 10^{-4} \qquad ''$$

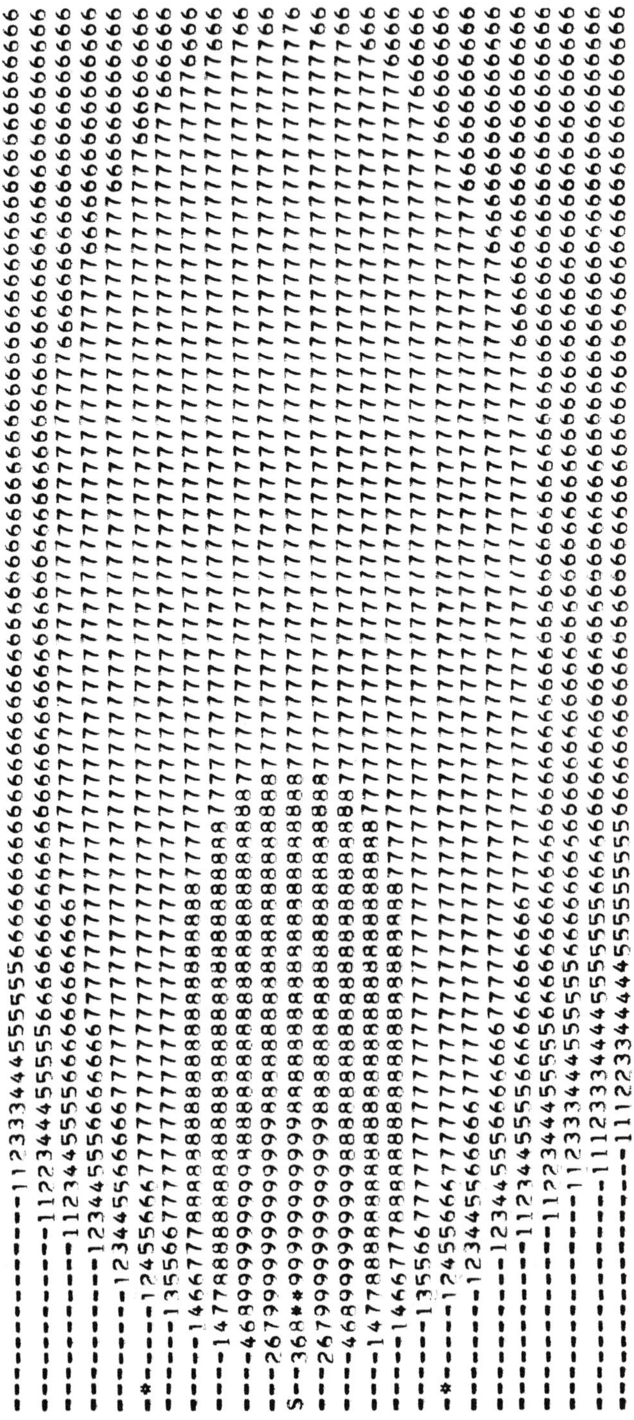

Figure 8.6. Dispersion plume under stability condition A calculated on a 100-meter grid for the stack conditions described in the text.

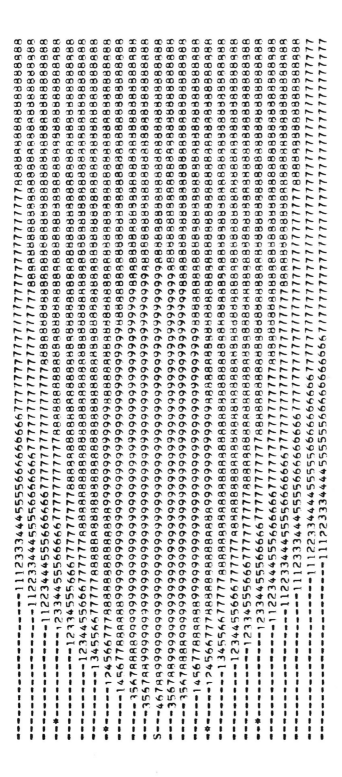

Figure 8.7. Dispersion plume under stability condition B calculated on a 100-meter grid for the stack conditions described in the text.

Figure 8.8. Dispersion plume under stability condition C calculated on a 100-meter grid for the stack conditions described in the text.

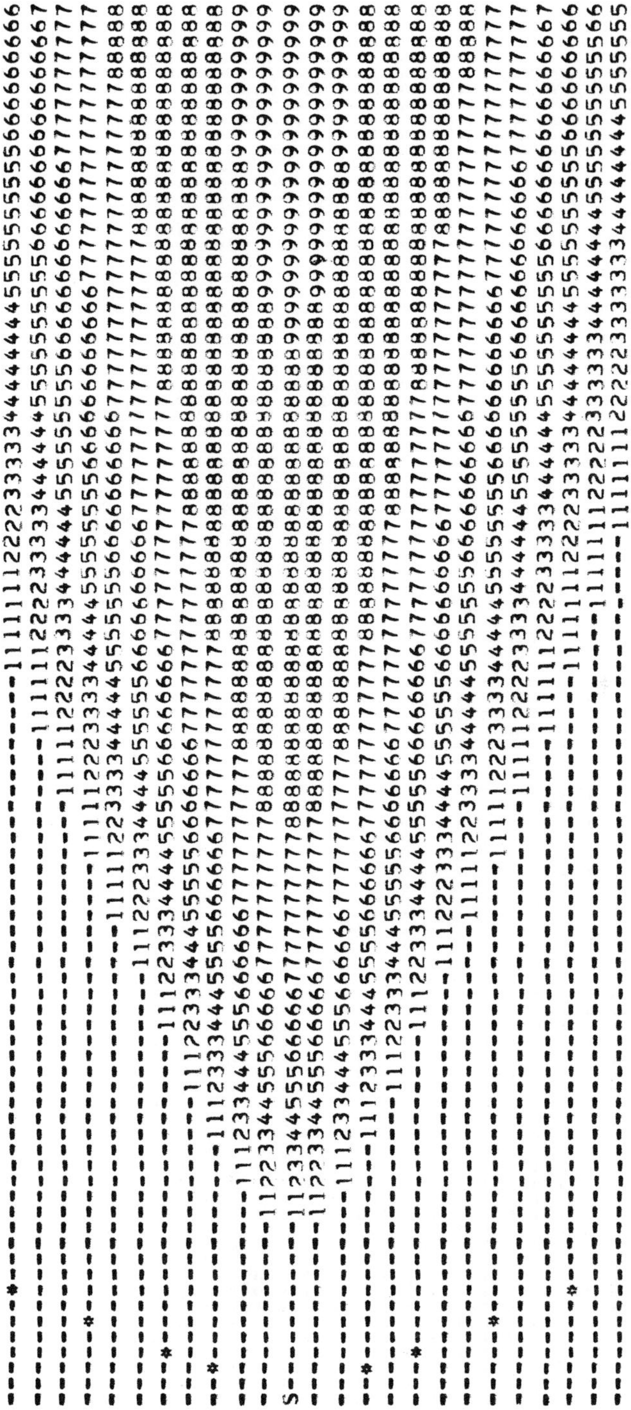

Figure 8.9. Dispersion plume under stability condition D calculated on a 100-meter grid for the stack conditions described in the text.

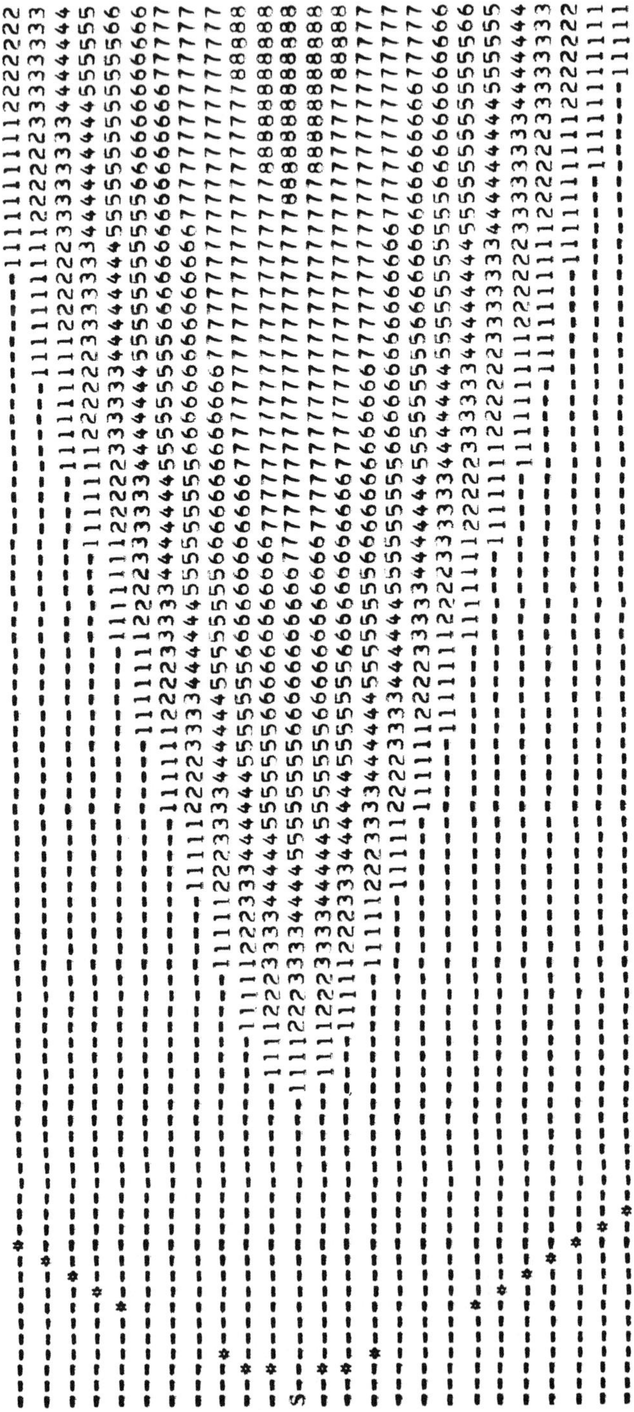

Figure 8.10. Dispersion plume under stability condition E calculated on a 100-meter grid for the stack conditions described in the text.

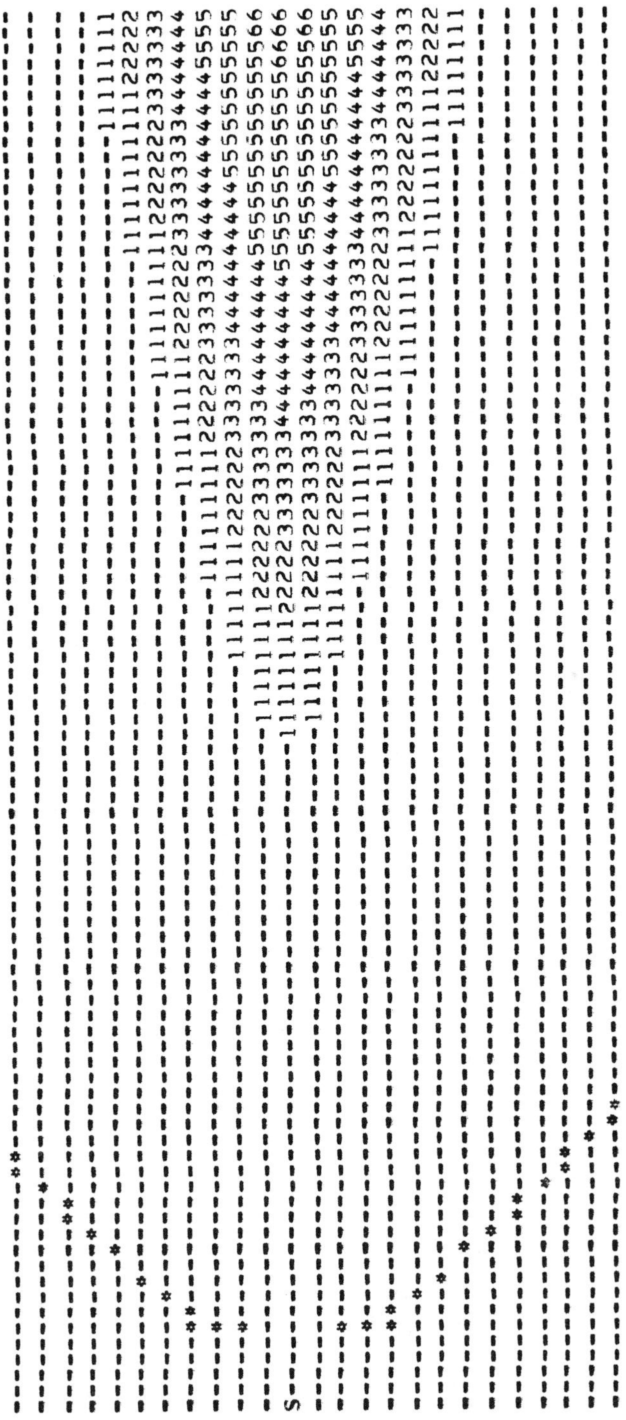

Figure 8.11. Dispersion plume under stability condition F calculated on a 100-meter grid for the stack conditions described in the text.

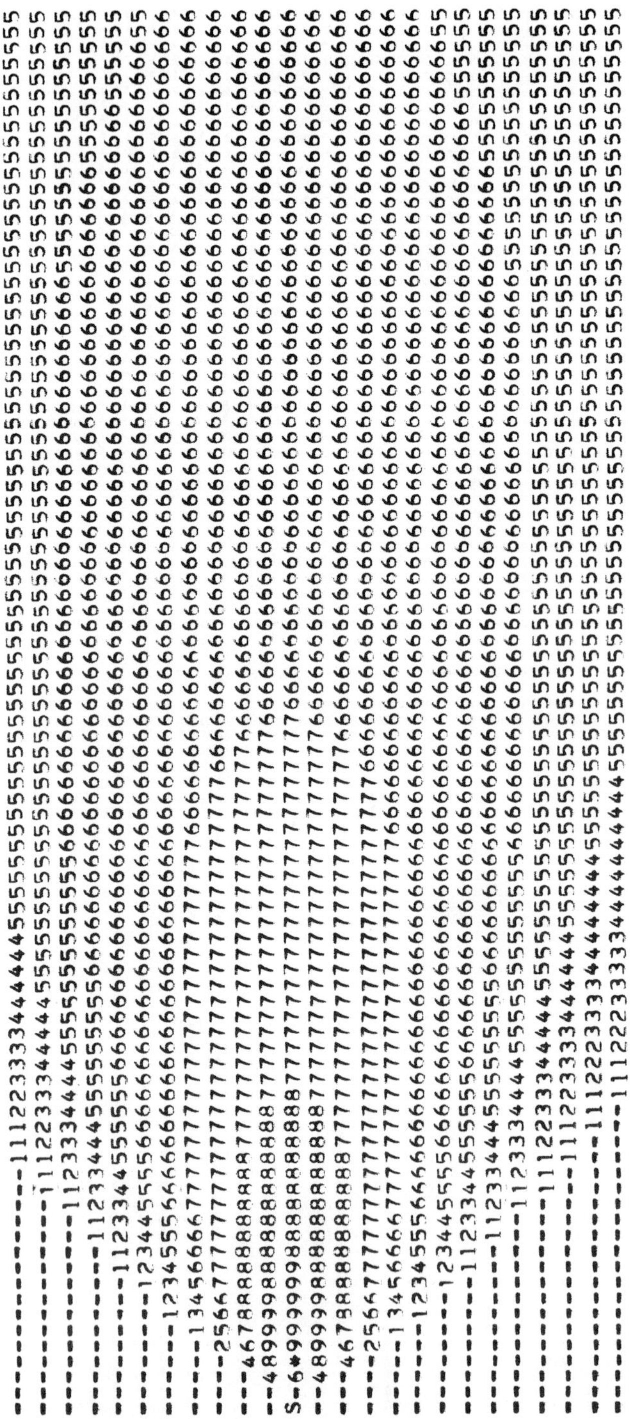

Figure 8.12. Dispersion plume under stability condition A calculated on a 200-meter grid for the stack conditions described in the text. This plume should be compared with that presented in Figure 8.6.

Figure 8.13. Dispersion plume under stability condition A calculated on a 300-meter grid for the stack conditions described in the text.

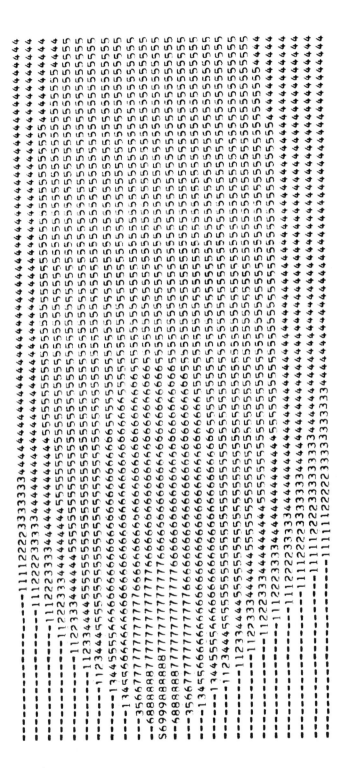

Figure 8.14. Dispersion plume under stability condition A calculated on a 400-meter grid for the stack conditions described in the text.

Figure 8.15. Dispersion plume under stability condition A calculated on a 500-meter grid for the stack conditions described in the text.

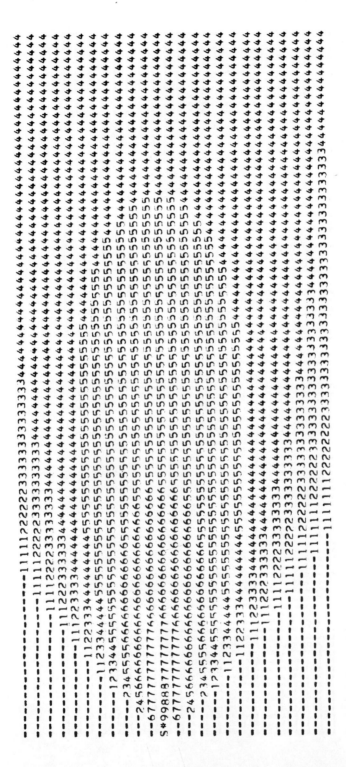

Figure 8.16. Dispersion plume under stability condition A calculated on a 600-meter grid for the stack conditions described in the text.

Figure 8.17. Dispersion plume under stability condition A calculated on a 700-meter grid for the stack conditions described in the text.

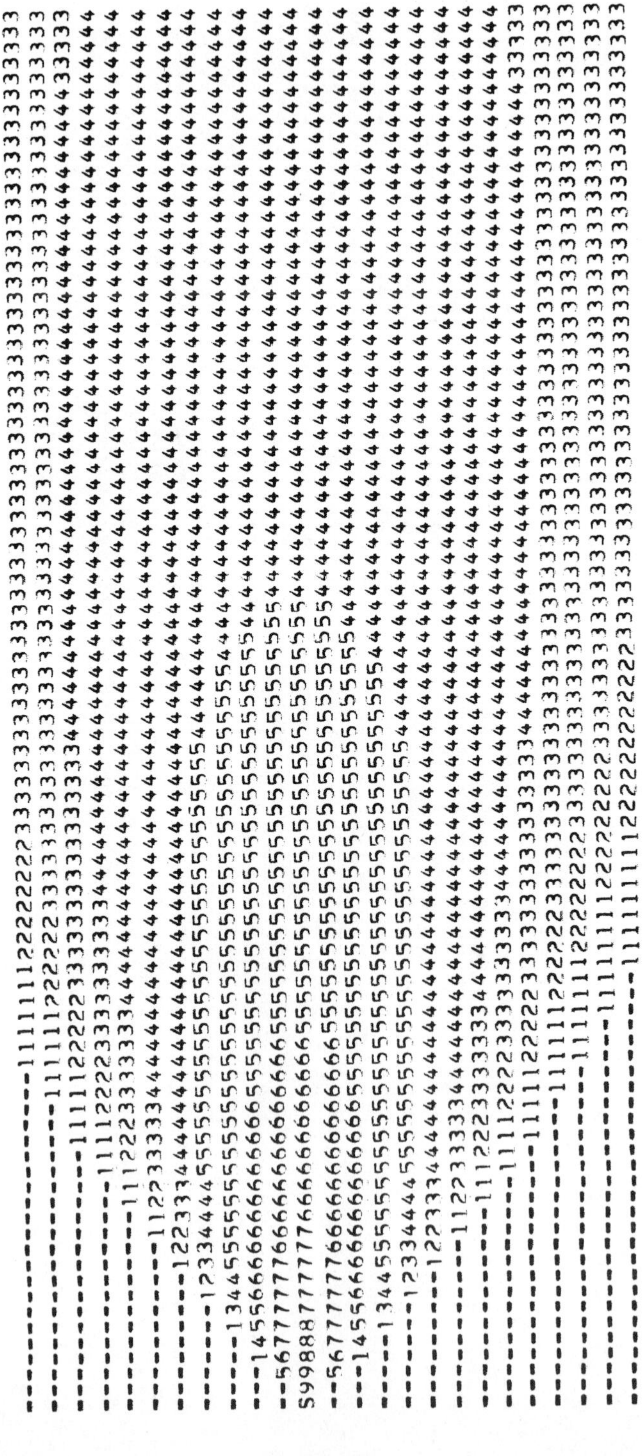

Figure 8.18. Dispersion plume under stability condition A calculated on an 800-meter grid for the stack conditions described in the text.

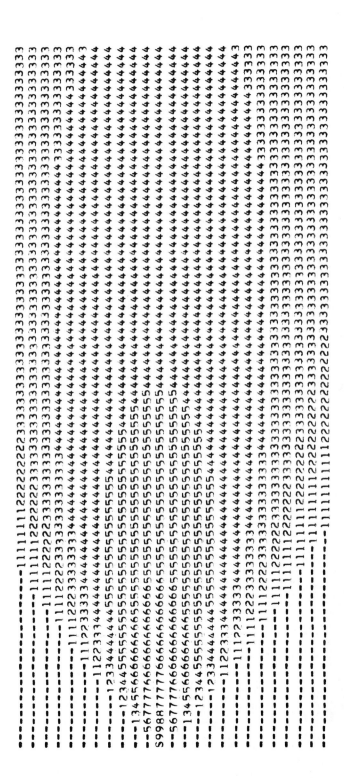

Figure 8.19. Dispersion plume under stability condition A calculated on a 900-meter grid for the stack conditions described in the text.

Figure 8.20. Dispersion plume under stability condition A calculated on a 1 km grid for the stack conditions described in the text.

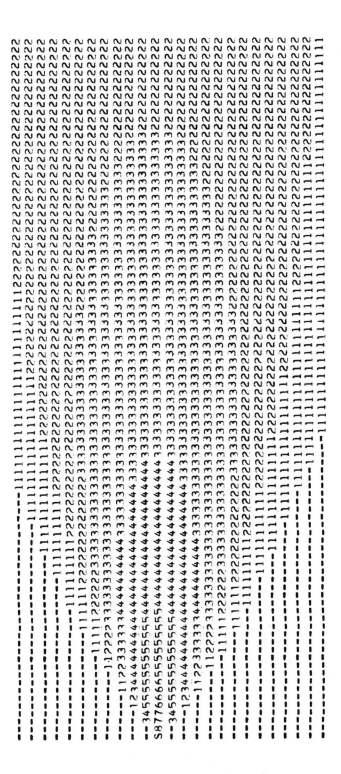

Figure 8.21. Dispersion plume under stability condition A calculated on a 2.5 km grid for the stack conditions described in the text.

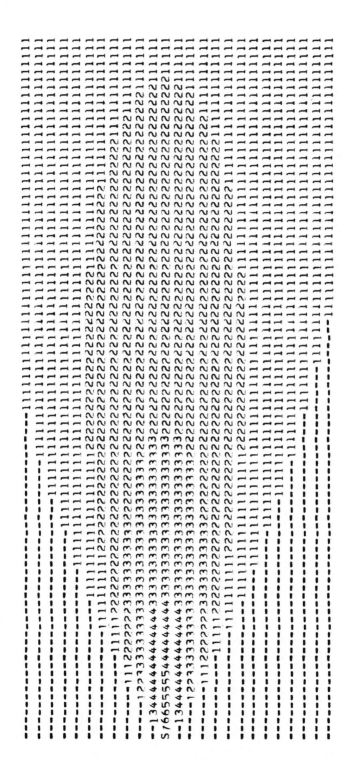

Figure 8.22.   Dispersion plume under stability condition A calculated on a 5 km grid for the stack conditions described in the text.

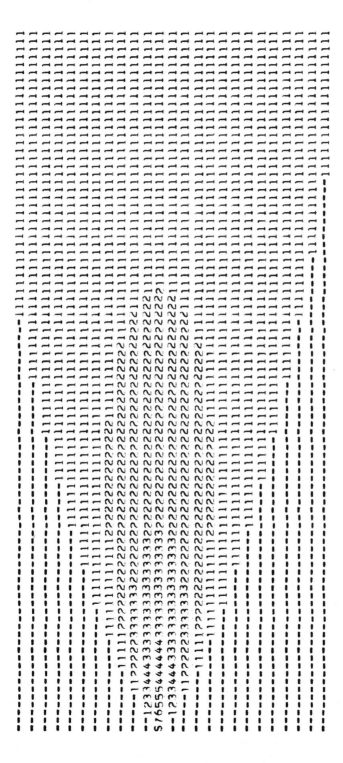

Figure 8.23. Dispersion plume under stability condition A calculated on a 7.5 km grid for the stack conditions described in the text.

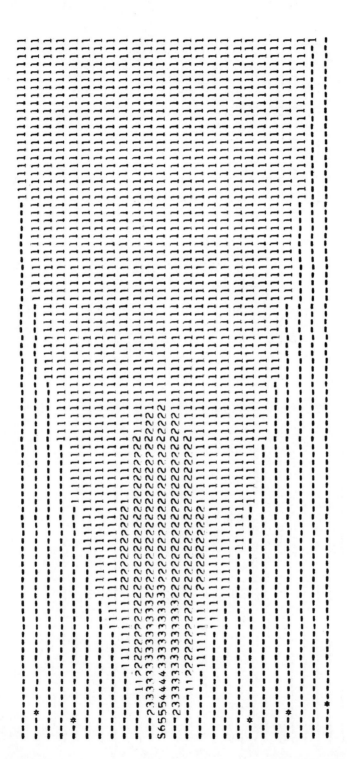

Figure 8.24. Dispersion plume under stability condition A calculated on a 10 km grid for the stack conditions described in the text.

   The contours indicated in Figures 8.6 through 8.11
illustrate the effect of varying stability under uniform
wind conditions, pointing out that large eddy transport
intensities (Case A, Figure 8.6) produce high surface pollution concentrations in nearby downwind locations while
less intense eddy transport moves the problem downstream
away from the pollution source (Case F, Figure 8.11).
   The detailed look at the plume provided by Figures
8.6-8.11 is then expanded for Case A (Figure 8.6) from the
100-meter grid size used in these first six illustrations to
200, 300, 400, 500, 600, 700, 800, 900 meter grids and 1,
2.5, 5, 7.5 and 10 kilometer grids in Figures 8.12-8.24,
respectively. As should be expected, the concentration is
decreased in strong gradient areas as the grid size is increased since the value being reported (q') is the average
concentration in that particular grid volume of unit depth.
   The examples presented here simply illustrate the
problem. The Computation Center at the University of Texas
at El Paso is currently programed and able to carry out
such calculations for this extremely simplified model for
any number of sources and provide an integrated presentation of the pollution field which would result.

CONCLUSIONS

The state of the atmosphere has a strong bearing on the ecology of the El Paso area. The supply of data available for application to study of ecological problems has been surveyed here, and models of the El Paso mesoscale meteorological structure have been derived from these data. It is concluded that currently available data are inadequate to provide the sophisticated atmospheric inputs required by the modern technological industry of El Paso.

A mesoscale meteorological observational network, data processing and presentation system has been outlined which is aimed at providing the atmospheric data which will be required in the immediate future. This mesomet network has been designed to be constructed and operated in conjunction with the public school system, an approach which is expected to yield improvements in the system through application of special talents and through education of the next generation of leaders relative to the merits of this data.

Examples of current practices relative to introduction of atmospheric variables into ecological problems have been presented. Use of single-valued parameters to represent the atmosphere in establishing and evaluating the ecological situation is clearly unacceptable to an advanced society in an environment which has the complexity of the El Paso region. Initiation of a mesomet system now is essential for the future health of El Paso.

Additional development will be required to integrate the information provided by such a mesomet system into calculations of their impact on the local ecology. Resources are available to accomplish this development, awaiting only the signal that use will be made of the results obtained. An aggressive approach to the general ecological problem is clearly in the best self interest of the community, and provision of an adequate source of atmospheric data is basic to the success of any environmental effort.

Evaluation of the impact of ecological factors on the technology of El Paso can best be carried out through use of the grid system presented here. Atmospheric influences on air pollution, fires, explosions, disease and other ecological processes require a more detailed analysis than has been applied in the past. This is particularly true for El Paso in light of our complex meteorological situation.

REFERENCES

Ballard, H. N., 1967. A Guide to Stratospheric Temperature and Wind Measurements. COSPAR Technique Manual Series, Secretariat, 55 Boulevard Malesherbes, Paris 8e, France, 117 p.

Ballard, H. N., 1970. A Cryogenic Sampler for the Determination of Stratospheric Composition. In SPACE TECHNOLOGY AND EARTH PROBLEMS, Vol. 23, Science and Technology Series, American Astronautical Society, Box 746, Tarzana, California 91356.

Beyers, N. J., 1969. Radar Chaff as a Wind Sensor. In STRATOSPHERIC CIRCULATION, pp 89-96, edited by W. L. Webb, Academic Press, Inc., New York, 600 p.

Beyers, N. J., and B. T. Miers, 1965. Diurnal Temperature Change in the Atmosphere between 30 and 60 Kilometers over White Sands Missile Range. Journal of the Atmospheric Sciences, 22, 3, 262-266.

Beyers, N. J., and B. T. Miers, 1968. A Tidal Experiment in the Equatorial Stratosphere over Ascension Island. Journal of the Atmospheric Sciences, 25, 1, 155-159.

Beyers, N. J., B. T. Miers and R. J. Reed, 1966. Diurnal Tidal Motions Near the Stratopause during 48 Hours at White Sands Missile Range. Journal of the Atmospheric Sciences, 23, 3, 325-333.

Booker, H. G., 1956. Turbulence in the Ionosphere with Applications to Meteor Trails, Radio Star Scintillation, Auroral Radar Echoes, and other Phenomena. Journal of Geophysical Research, 61, 673-705.

Chalmers, J. A., 1967. ATMOSPHERIC ELECTRICITY. Pergamon Press, Oxford, 515 p.

Chapman, S., and J. Bartels, 1940. GEOMAGNETISM, Vols. 1 & 2. Clarendon Press, Oxford, 1064 p.

Chapman, S., 1951. Photochemical Processes in the Upper Atmosphere and Resultant Composition. In COMPENDIUM OF METEOROLOGY, pp. 262-274, edited by T. F. Malone, American Meteorological Society, Boston, Mass., 1311 p.

Cole, R. K., Jr., and E. T. Pierce, 1965. Electrification in the Earth's Atmosphere for Altitudes between 0 and 100 Kilometers. Journal of Geophysical Research, 70, 2735-2749.

Dean, E. A., 1959. Absorption of Low Frequency Sound in a Homogeneous Atmosphere. Texas Western College, El Paso, Schellenger Research Laboratory, Contract DA-29-040-ORD-1237, 86 p.

Duff, A. A., 1971. White Sands Missile Range Meteor Trail Radar Design. In THERMOSPHERIC CIRCULATION, pp. 355-367, edited by W. L. Webb, MIT Press, Cambridge, 600 p.

Eddy, A., 1969. A Meteorologically Oriented Computer Model of an Ecosystem. Transactions of the New York Academy of Sciences, Series II, 31, 6, 618-628.

Fejer, J. A., 1965. Motions of Ionization. In PHYSICS OF THE EARTH'S UPPER ATMOSPHERE, pp. 157-175, editors C. O. Hines, I. Paghis, T. R. Hartz and J. A. Fejer, Prentice-Hall, Inc., Englewood Cliffs, New Jersey, 434 p.

Glass, R. I., 1964. Missile Meteorology Tower. Bulletin of the American Meteorological Society, 45, 9, 601-602.

Glass, R. I., R. D. Reynolds and R. L. Lamberth, 1968. A High Resolution Continuous Pressure Sensor Modification for Radiosondes. Journal of Applied Meteorology, 7, 1, 141-144.

Hanson, W. B., 1965. Structure of the Ionosphere. In SATELLITE ENVIRONMENT HANDBOOK, 2nd Edition, pp. 27-46, edited by F. S. Johnson, Stanford University Press, 193 p.

Heimbach, J. A., 1970. Pulmonary Emphysema and Lung Cancer as Related to Air Pollution in Altoona, Pennsylvania. MS Thesis, Meteorology Department, University of Oklahoma, Norman, Oklahoma.

Hinteregger, H. E., L. A. Hall and G. Schmidtke, 1965. Solar XUV Radiation and Neutral Particle Distribution in July 1963 Thermosphere. In SPACE RESEARCH, Vol. V, pp 1175-1190, edited by D. G. King-Hele, P. Muller and G. Righini, John Wiley and Sons, Inc., 1248 p.

Hoidale, Marjorie M., B. J. Gee and M. A. Seagraves, 1968a. Atmospheric Structure, White Sands Missile Range, New Mexico, Part 3, Upper Air Data: Stallion Site. Atmospheric Sciences Laboratory DR 323, U. S. Army Electronics Command, White Sands Missile Range, New Mexico.

Hoidale, Marjorie M., B. J. Gee and M. A. Seagraves, 1968b. Atmospheric Structure, White Sands Missile Range, New Mexico, Part 3, Upper Air Data: Holloman. Atmospheric Sciences Laboratory DR 321, U. S. Army Electronics Command, White Sands Missile Range, New Mexico.

Hoidale, Marjorie M., B. J. Gee and M. A. Seagraves, 1968c. Atmospheric Structure, White Sands Missile Range, New Mexico, Part 3, Upper Air Data: Apache. Atmospheric Sciences Laboratory DR 322, U. S. Army Electronics Command, White Sands Missile Range, New Mexico.

Hoidale, Marjorie M., B. J. Gee and M. A. Seagraves, 1968d.
Atmospheric Structure, White Sands Missile Range, New Mexico,
Part 3, Upper Air Data: White Sands Desert. Atmospheric
Sciences Laboratory DR 327, U. S. Army Electronics Command,
White Sands Missile Range, New Mexico.

Hoidale, Marjorie M., B. J. Gee and M. A. Seagraves, 1969a.
Atmospheric Structure, White Sands Missile Range, New Mexico,
Part 3, Upper Air Data: Small Missile Range. Atmospheric
Sciences Laboratory DR 324, U. S. Army Electronics Command,
White Sands Missile Range, New Mexico.

Hoidale, Marjorie M., B. J. Gee and M. A. Seagraves, 1969b.
Atmospheric Structure, White Sands Missile Range, New Mexico,
Part 3, Upper Air Data: Jallen. Atmospheric Sciences Lab-
oratory DR 325, U. S. Army Electronics Command, White Sands
Missile Range, New Mexico.

Hoidale, Marjorie M., B. J. Gee and M. A. Seagraves, 1969c.
Atmospheric Structure, Utah Launch Complex, Green River,
Utah, Part 3, Upper Air Data: Green River, Utah. Atmo-
spheric Sciences Laboratory DR 326, U. S. Army Electronics
Command, White Sands Missile Range, New Mexico.

Holzworth, G. C., 1964. Estimates of Mean Maximum Mixing
Depths in the Contiguous United States. Monthly Weather
Review, 92, 5, 235-242.

Inman, R. L., 1970. Operational Objective Analysis Systems
at the National Severe Storms Forecast Center. National
Severe Storms Laboratory, Technical Circular No. 10, Norman,
Oklahoma.

Johnson, F. S., 1966. Turbopause Processes and Effects.
In SPACE RESEARCH, Vol. VII, Vols. 1 & 2, pp. 262-269,
edited by R. L. Smith-Rose, North-Holland Publishing Company,
Amsterdam, 1479 p.

Jones, W. L., 1969. Atmospheric Internal Gravity Waves and
Tides. In STRATOSPHERIC CIRCULATION, pp. 469-482, edited
by W. L. Webb, Academic Press, Inc., New York, 600 p.

Lamberth, R. L., and R. D. Reynolds, 1965. Mountain Lee
Waves at White Sands Missile Range. Bulletin of the American Meteorological Society, 46, 10, 634-636.

Leovy, C., 1964. Radiative Equilibrium of the Mesosphere.
Journal of the Atmospheric Sciences, 21, 238-248.

Lettau, H., 1951. Diffusion in the Upper Atmosphere. In
COMPENDIUM OF METEOROLOGY, pp. 320-333, edited by T. E.
Malone, American Meteorological Society, Boston, Mass.,
1311 p.

Lindzen, R. S., 1967. Thermally Driven Diurnal Tides in the Atmosphere. Quarterly Journal of the Royal Meteorological Society, 93, 18-42.

Murrow, H. N., 1969. Application of Decelerators to Atmospheric Probes. In STRATOSPHERIC CIRCULATION, pp. 97-113, edited by W. L. Webb, Academic Press, Inc., New York, 600 p.

Miers, B. T., 1965. Wind Oscillations between 30 and 60 Kilometers over White Sands Missile Range, New Mexico. Journal of the Atmospheric Sciences, 22, 4, 382-387.

Newell, H. E., 1959. SOUNDING ROCKETS. McGraw-Hill Book Company, Inc., New York, 334 p.

Pasquill, F., 1962. ATMOSPHERIC DIFFUSION. D. Van Nostrand Company, Ltd., London, 297 p.

Randhawa, J. S., 1967. Ozonesonde for Rocket Flight. Nature, 213, 53.

Randhawa, J. S., 1968. Mesospheric Ozone Measurements during a Solar Eclipse. Journal of Geophysical Research, 73, 2, 493-495.

Randhawa, J. S., 1969a. Ozone Measurements from a Stable Platform near the Stratopause Level. Journal of Geophysical Research, 74, 18, 4588-4590.

Randhawa, J. S., 1969b. Chemiluminescent Ozone Measurements. In STRATOSPHERIC CIRCULATION, pp 175-182, edited by W. L. Webb, Academic Press, Inc., New York, 600 p.

Reiter, E. R., 1963. JET-STREAM METEOROLOGY. University of Chicago Press, 515 p.

Reynolds, R. D., and R. L. Lamberth, 1966. Ambient Temperature Measurements from Radiosondes Flown on Constant Level Balloons. Journal of Applied Meteorology, 5, 3, 304-307.

Reynolds, R. D., R. L. Lamberth and M. G. Wurtele, 1968. Investigation of a Complex Mountain Wave Situation. Journal of Applied Meteorology, 7, 3, 353-358.

Reynolds, R. D., 1969. Dichotomy of Mountain Lee Wave Theories and Field Test Measurements. Annalen der Meteorologie, N. F. Nr. 4, 234-239.

Sayre, A. N., and P. P. Livingston, 1945. Ground-Water Resources of the El Paso Area, Texas. United States Geological Survey Water Supply Paper 919, 190 p.

Scherhag, R., 1952. Die Explosionsartigen Stratosphärener-
warmungen des Spätwinters 1951/52. Berlin Deutland Wetter-
dienstes, U. S. Zone, 38, 51-63.

Slade, D. H., 1968. METEOROLOGY and ATOMIC ENERGY 1968.
USAEC Division of Technical Information Extension, Oak
Ridge, Tennessee, TID-24190, 445 p.

Sonnichsen, C. L., 1968. PASS OF THE NORTH. Texas Western
Press, University of Texas at El Paso, 467 p.

Strain, W. S., 1966. Blancan Mammalian Fauna and Pleistocene
Formations, Hudspeth County, Texas. Bulletin of the Texas
Memorial Museum, University of Texas at Austin, 55 p.

Taft, P. H., and Marjorie M. Hoidale, 1968. White Sands
Missile Range Climatography, No. 2, Holloman. Atmospheric
Sciences Laboratory DR 379, U. S. Army Electronics Command,
White Sands Missile Range, New Mexico (AD 844 658).

Taft, P. H., and Marjorie M. Hoidale, 1969a. White Sands
Missile Range Climatography, No. 5, Stallion Site, WSMR.
Atmospheric Sciences Laboratory DR 399, U. S. Army Elec-
tronics Command, White Sands Missile Range, New Mexico
(AD856 240).

Taft, P. H., and Marjorie M. Hoidale, 1969b. White Sands
Missile Range Climatography, No. 3, Apache. Atmospheric
Sciences Laboratory DR 387, U. S. Army Electronics Command,
White Sands Missile Range, New Mexico (AD 848 856).

Taft, P. H., and Marjorie M. Hoidale, 1969c. White Sands
Missile Range Climatography, No. 4, White Sands Desert.
Atmospheric Sciences Laboratory DR 397, U. S. Army Elec-
tronics Command, White Sands Missile Range, New Mexico
(AD 856 360).

Taft, P. H., and Marjorie M. Hoidale, 1969d. White Sands
Missile Range Climatography, No. 6, Small Missile Range.
Atmospheric Sciences Laboratory DR 400, U. S. Army Elec-
tronics Command, White Sands Missile Range, New Mexico
(AD 860 148).

Taft, P. H., and Marjorie M. Hoidale, 1969e. White Sands
Missile Range Climatography, No. 7, Jallen. Atmospheric
Sciences Laboratory DR 416, U. S. Army Electronics Command,
White Sands Missile Range, New Mexico (AD 857 661).

Taft, P. H., and Marjorie M. Hoidale, 1969f. Utah Launch
Complex Climatography, Green River, Utah. Atmospheric
Sciences Laboratory DR 401, U. S. Army Electronics Command,
White Sands Missile Range, New Mexico, (AD 860 131).

Webb, W. L., 1962. Detailed Acoustic Structure above the Tropopause. Journal of Applied Meteorology, 1, 2, 229-236.

Webb, W. L., 1965. Stratospheric Solar Response. Journal of the Atmospheric Sciences, 21, 6, 582-591.

Webb, W. L., 1966a. STRUCTURE OF THE STRATOSPHERE AND MESO-SPHERE. Academic Press, Inc., New York, 382 p.

Webb, W. L., 1966b. Stratospheric Tidal Circulations. Reviews of Geophysics, 4, 3, 363-375.

Webb, W. L., 1968a. Atmospheric Diurnal Electrical Structure. SPACE RESEARCH VIII, pp 896-906, edited by A. P. Mitra, L. G. Jacchia and W. S. Newman, North Holland Publishing Co., Amsterdam, 1096 p.

Webb, W. L., 1968b. Source of Atmospheric Electrification. Journal of Geophysical Research, 73, 16, 5061-5071.

Webb, W. L., 1969a. STRATOSPHERIC CIRCULATION, editor. Academic Press, Inc., New York, 600 p.

Webb, W. L., 1969b. Global Electrical Structure. PLANETARY ELECTRODYNAMICS, Vol. II, edited by S. C. Coroniti and J. Hughes, Gordon and Breach Science Publishers, New York, 583 p.

Webb, W. L., 1970a. The Cold Earth. In SPACE TECHNOLOGY and EARTH PROBLEMS, Volume 23, pp. 85-96, Science and Technology Series, American Astronautical Society, P. O. Box 746, Tarzana, California 91256.

Webb, W. L., 1970b. Global Electrical Currents. Pure and Applied Geophysics, 82, 87-106.

Webb, W. L., 1971. THERMOSPHERIC CIRCULATION, editor. MIT Press, Cambridge, Mass., 372 p.

Webb, W. L., J. W. Coffman and G. Q. Clark, 1959. A High Altitude Acoustic Sensing System. Signal Missile Support Agency Report No. 28, White Sands Missile Range, New Mexico, 38 p.

Webb, W. L., W. E. Hubert, R. L. Miller and J. F. Spurling, 1961. The First Meteorological Rocket Network. Bulletin of the American Meteorological Society, 42, 7, 482-494.

Webb, W. L., W. I. Christensen, E. P. Varner and J. F. Spurling, 1962. Inter-Range Instrumentation Group Participation in the Meteorological Rocket Network. Bulletin of the American Meteorological Society, 43, 12, 640-649.

Webb, W. L., and K. R. Jenkins, 1962. Sonic Structure of the Mesosphere. Journal of the Acoustical Society of America, 34, 2, 193-211.

Webb, W. L., J. Giraytys, H. B. Tolefson, R. C. Forsberg, R. I. Vick, O. H. Daniel and L. R. Tucker, 1966. Meteorological Rocket Network Probing of the Stratosphere and Mesosphere. Bulletin of the American Meteorological Society, 47, 10, 788-799.

Wright, J. B., 1969. The Robin Falling Sphere. In STRATOSPHERIC CIRCULATION, pp. 115-139, edited by W. L. Webb, Academic Press, Inc., New York, 600 p.

Appendix I Fortran program used to interpolate and extrap-
olate mesonet data onto a computational grid. This program follows that developed by Inman (1970) for the National
Severe Storms Forecast Center mesoscale network.

```
      PROGRAM EPH7P001
C
C     INTERPOLATE STATION ELEVATIONS ONTO GRID POINTS
C
      COMMON NI,NJ,NIM1,NJM1,IDX,IDY,IXO,IYO,IXFO,IYFO,
     1      IXGP(40),IYGP(40),NOS,NOP,IR(6),ICA(23),IZH
     2      (23),ICALL(23),JCALL(23),IZGE(40,40),IX(23),
     3      IY(23),IU(23),IV(23),IM(23),JX(40,40),JY(40,40),
     4      JCA(23),IZJ(23),LU(23),LV(23)
      DIMENSION ITH(6),IZX(23),IZGX(40,40),IZGY(40,40)
      DIMENSION LB(5)
      COMMON/DATA/KALP(16)
      DATA ((KALP(I),I=1,16)=1H ,1HX,1H ,1HX,1H ,1HX,1H,
     *      1HX,1H ,1HX,1H,1HX,1H ,1HX,1H ,1HX)
      LB(1)=4H  C $ LB(2)=4HONTO $ LB(3)=4HUR I $LB(4)=
     4HNTER  LB(5)=4HVAL
    5 READ 1000, NOS,NI,NJ,NOP,IXO,IYO,IDX,IDY,JJJ,IR,ITH,
     *      (IX(I),I=1,NOS),(IY(I),I=1,NOS),(IZX(I),I=1,
     *      NOS),IFLAG
 1020 FORMAT (10X,6I8)
 1000 FORMAT (9I5/12I5,3(/12I6/11I6),10X,A4)
      PRINT1000, NOS,NI,NJ,NOP,IXO,IYO,IDX,IDY,JJJ,IR,ITH,
     *      (IX(I),I=1,NOS),(IY(I),I=1,NOS),(IZX(I),I=1,
            NOS) ,IFLAG
      DO 10  I = 1,NI
   10 IXGP(I) = I*IDX
      DO 20 I = 1,NJ
   20 IYGP(I) = I*IDY
      NIM1 = NI - 1   $ NJM1 = NJ - 1
      CALL  OBAN(IZX,IZGY,JJJ,ITH)
      PRINT 1010
      PRINT 1020,((I,J,IXGP(I),IYGP(J),IZGY(I,J),IZGE(I,J)
     *      ,I=1,NI),J=1,NJ)
C     TRANSPOSE THE MATRIX AND REVERSE THE ORDER OF THE
      ROWS.
C     THIS WILL ALLOW CONTUR TO PRINT AN UPRIGHT NIXNJ GRID.
      DO 30 I = 1,NI
      DO 30 J = 1,NJ
      II = NI+1 - I
   30 IZGX(II,J) = IZGE(J,I)
      DO 40 I = 1,NI
      DO 40 J = 1,NJ
      II = NI+1 - I
   40 IZGE(II,J) = IZGY(J,I)
 1010 FORMAT (1H1)
C     PRINT UNSMOOTHED CONTOUR MAPS.
      PRINT 1010
      CALL CONTUR(IZGE,NI,26,0,100,4H100M,LB)
      PRINT 1010
```

```
       CALL CONTUR(IZGE(1,26),NI,15,0,100,4H100M,LB)
       PRINT 1010
       CALL CONTUR (IZGE,NI,26,0,50,4H 50M,LB)
       PRINT 1010
       CALL CONTUR(IZGE(1,26),NI,15,0,50,4H 50M,LB)
       PRINT 1010
       CALL CONTUR(IZGE,NI,26,0,25,4H 25M,LB)
       PRINT 1010
       CALL CONTUR(IZGE(1,26),NI,15,0,25,4H 25M,LB)
       PRINT 1010
       CALL CONTUR(IZGE(1,1),NI,26,0,10,4H 10M,LB)
       PRINT 1010
       CALL CONTUR(IZGE(1,26),NI,15,0,10,4H 10M,LB)
C      PRINT SMOOTHED CONTOUR MAPS.
       PRINT 1010
       CALL CONTUR(IZGX,NI,26,0,100,4H100M,LB)
       PRINT 1010
       CALL CONTUR(IZGX(1,26),NI,15,0,100,4H100M,LB)
       PRINT 1010
       CALL CONTUR (IZGX,NI,26,0,50,4H 50M,LB)
       PRINT 1010
       CALL CONTUR(IZGX(1,26),NI,15,0,50,4H 50M,LB)
       PRINT 1010
       CALL CONTUR(IZGX,NI,26,0,25,4H 25M,LB)
       PRINT 1010
       CALL CONTUR(IZGX(1,26),NI,15,0,25,4H 25M,LB)
       PRINT 1010
       CALL CONTUR(IZGX(1,1),NI,26,0,10,4H 10M,LB)
       PRINT 1010
       CALL CONTUR (IZGX(1,26),NI,15,0,10,4H 10M,LB)
       IF (IFLAG.EQ.4HSTOP) RETURN
       GO TO 5
       END

       SUBROUTINE OBAN(IZX,IZGX,JJJ,IOUT)
C      ...ONE FIELD SCHEME WITH MISSING DATA PROVISION
C      ...JJJ=-1, STATION WIND VELOCITY USED
C      ...JJJ= 0, SKIP SPECIAL FORMULATION
C      ...JJJ= 1, ANALYZED WIND VELOCITY AT GRID POINTS USED
C      ...MISSING DATA INDICATED BY 32767 in IZX,IU,IV,AND IM
C      ...IZX HOLDS OBS AND IS UNCHANGED
C      ...IZX SHOULD HAVE MAGNITUDE LESS THAN 1000
C      ...IZH HOLDS DEVIATION (OBSERVATION MINUS ANALYZED
C         VALUE)
C      ...ICA HOLDS ANALYZED VALUE OF EACH STATION
C      ...IZGX HOLDS ANALYZED VALUE AT GRID POINTS
C      ...IX X-COORDINATE OF OBS STN (MEASURED IN UNITS OF
C      1/100 INCH)
C      ...IY Y-COORDINATE OF OBS STN (MEASURED IN UNITS OF
C      1/100 INCH)
C      ...IU HOLDS X-COMPONENT OF WIND VELOCITY
C      ...IV HOLDS Y-COMPONENT OF WIND VELOCITY
C      CCC+M HOLDS SQUARE OF WIND SPEED
C      ...JX AND JY HOLD ANALYZED U,V COMP. OF WIND (NEEDED
```

```
        ONLY IF JJJ=1)
C       ...IOUT HOLDS TOOSS-OUT CRITERIA
        COMMON NI,NJ,NIM1,NJM1,IDX,IDY,IXO,IYO,IXFO,IYFO,
     1  IXGP(40),IYGP(40),NOS,NOP,IR(6),ICA(23),IZH(23),
     2  ICALL(23),JCALL(23),IZGE(40,40),IX(23),IY(23),IU(23),
     3  IV(23),IM(23),JX(40,40),JY(40,40),JCA(23),IZJ(23),
     4  LU(23),LV(23)
        DIMENSION IZX(23),IZGX(40,40),IOUT(6)
        IPHASE = 0
        DO 2 I=1,NI
        DO 2 J=1,NJ
     2  IZGX(I,J)=0
        NNS=2
C    A CONSTANT RADIUS OF NR IS USED FOR STN OUTSIDE GRID
     ON ALL SCANS.
        NR=300
        NR2=NR*NR
C    LOCATES GRID SQUARE CONTAINING OBSERVATION STATION
        DO 50 K=1, NOS
        IF(IX(K)-IXO)16,17,18
     18 IF (IX(K)-IXFO)17,16,16
     17 IF   (IY(K)-IYO) 16,15,19
     19 IF(IY(K)-IYFO)15,16,16
C       ...STATION IS OUTSIDE GRID
     16 ICALL(K)=999
C       GO TO 50
     15 ICALL(K)=1+(IX(K)-IXO)/IDX
        JCALL(K)=1+(IY(K)-IYO)/IDY
     50 CONTINUE
C    INITIALIZATION OF ARRAYS FOR GRID
        DO 4 K=1,NOS
        ICA(K)=0
     4  IZH(K)=IZX(K)
        DO 199 L=1,NOP
        KR=IR(L)
        IR2=KR*KR
C    ON SCAN 1 INITIAL GUESS IS PRODUCED
        IF(L-1) 123,23,123
C    INTERPOLATION OF ANALYSIS TO OBSERVATION LOCATION
     (SCANS 2,3 and 4)
    123 DO 22 K=1,NOS
        IF(IZX(K)-32767)10022,22,22
  10022  IXK=IX(K)
        IYK=IY(K)
        IF(ICALL(K)-999)52,516,52
C       CALCULATE DEVIATION AT STATION WHEN STATION IS OUTSIDE
        GRID
    516 IA1=0
        IA2=0
        DO 524 I= 1,NOS
        IF(IZX(I)-32767)10524,524,10524
  10524  M1=IXK-IX(I)
        IF (IABS(M1)-NR) 525,524,524
    525 M2=IYK-IY(I)
```

```
         IF (IABS(M2)-NR) 526,524,524
   526 M3 = M1*M1+M2*M2
         IB1=NR2-M3
         IF(IB1)524,524,536
C    SPECIAL FORMULATION (WEIGHT FUNCTION MODIFIED ACCORD-
     ING TO WIND VEL)
   536 IF(JJJ)636,736,836
   636 IF(M3)3736,736,3736
  3736 IF(IM(I)-32767)5736,736,1736
  5736 IF(IM(I))1736,736,1736
  1736 D=IU(I)*M1+IV(I)*M2
         LR=NR*(1.+(D*D/(FLOAT(M3)*IM(I))))
         GO TO 438
   836 IF(M3)4736,736,4736
  4736 IF(IM(K)-32767)6736,736,2736
  6736 IF(IM(K))2736,736,2736
  2736 D=IU(K)*M1+IV(K)*M2
         LR=NR*(1.+(K*D/(FLOAT(M3)*IM(K))))
   438 LR2=LR*LR
         IB1=LR2-M3
         B2=LR2+M3
         GO TO 936
C    ...LAST OF SPECIAL FORMULATION
   736 B2=NR2+M3
   936 KW=(IB1/B2)*100
         IA1=IA1+KW*IZH(I)
         IA2=IA2+KW
   524 CONTINUE
         ICA(K)=(ICA(K)+IA1/IA2+IZX(K))/2
         GO TO 22
C    CALCULATION OF DEVIATION WHEN STATION IS WITHIN GRID
C    BILINEAR INTERPOLATION USING FOUR GRID POINTS SURROUND-
     ING THE STN.
    52 M=ICALL(K)
         N=JCALL(K)
         M1=IXK-IXGP(M)
         M2=IYK-IYGP(N)
         IZ1=IZGX(M,N)
         IZ2=IZGX(M,N+1)
         IZ4=IZGX(M+1,N)
         ICA(K)=IZ1+((M1*(IZ4-IZ1))/IDX+(M2*(IZ2-IZ1))/IDY+
     1 (((M1*M2)/IDX)*(IZGX(M+1,N+1)-IZ4+IZ1-IZ2))/IDY)
    22 CONTINUE
         DO 950 K=1,NOS
         IZH(K)=IZX(K)-ICA(K)
         IF(IZH(K)-32767)10948,950,10948
 10948 IF(IABS(IZH(K))-IOUT(L))950,950,10950
 10950 WRITE(61,10946) IX(K),IY(K),L,IZX(K),IZH(K),IOUT(L)
 10946 FORMAT(1H ,8HDATA AT ,2I10,2X,17HREJECTED ON PASS
     ,I1,2X,
     $    4HIZX=I7,2X,4HIZH=I7,2X,5HIOUT=I5)
         IZX(K)=32767
         IZH(K)=32767
         ICA(K)=0
```

```
      950 CONTINUE
C     CALCULATION OF CORRECTION TO GRID POINT VALUE
       23 DO 198 J=1,NJ
          IYK=IYGP(J)
          DO 198 I=1,NI
          JR=KR
          JR2=IR2
          IXK=IXGP(I)
          IF (JJJ) 1136,1136,1336
     1336 JU=JX(I,J)
          JV=JY(I,J)
          V=JU*JU+JV*JV
     1136 IA1=0
          IA2=0
          NN=0
          DO 24 K=1,NOS
          IF(IZX(K)-32767)10024,24,10024
    10024 M1=IX(K)-IXK
          IF(IABS(M1)-JR)25,24,24
       25 M2=IY(K)-IYK
          IF(IABS(M2)-JR)26,24,24
       26 M3=M1*M1+M2*M2
          IB1=JR2-M3
          IF(IB1)24,24,36
C     SPECIAL FORMULATION (WEIGHT FUNCTION MODIFIED ACCORD-
          ING TO WIND VEL)
       36 IF(JJJ)136,236,336
      336 IF(V)3236,236,3236
     3236 IF(M3)1236,236,1236
     1236 D=JU*M1+JV*M2
          LR=JR*(1.+(D*D/(V*M3)))
      138 LR2=LR*LR
          IB1=LR2-M3
          B2=LR2+M3
          GO TO 436
      136 IF(M3)4236,236,4236
     4236 IF(IM(K)-32767)5236,236,2236
     5236 IF(IM(K))2236,236,2236
     2236 D=IU(K)*M1+IV(K)*M2
          LR=JR*(1.+(D*D/(FLOAT(M3)*IM(K))))
          GO TO 138
C        ...LAST OF SPECIAL FORMULATION CARDS
      236 B2=JR2+M3
      436 KW=(IB1/B2)*100
          IA1=IA1+KW*IZH(K)
          IA2=IA2+KW
          NN=NN+1
       24 CONTINUE
C     TWO STN ARE REQUIRED WITHIN JR ON THE FIRST SCAN ONLY
          IF(NN-NNS)398,201,201
      201 IF(NN-1)198,202,200
C        NO STATIONS WITHIN JR - INCREASE JR AND TRY AGAIN
      398 JR=JR+IDX
          JR2=JR*JR
```

```
          GO TO 1136
    200 IF(IA2)398,398,27
    202 IA2=100
     27 IZGX(I,J)=IZGX(I,J)+IA1/IA2
   2000 FORMAT (12H END PHASE ,I2)
    198 CONTINUE
          NNS=0
          CALL SMOOTH(IZGX,IZGE)
          IPHASE = IPHASE + 1
C         WRITE (59,2000) IPHASE
    199 CONTINUE
          END
          SUBROUTINE SMOOTH(JW,IUX)
          DIMENSION JW(40,40),IUX(40,40)
          COMMON NI,NJ,NIM1,NJM1,IDX,IDY,IXO,IYO,IXFO,IYFO,
          IXGP(40),IYGP(40)
          DO 10 J=2,NJM1
          DO 10 I=2,NIM1
     10 IUX(I,J)=(4*JW(I,J)+JW(I-1,J)+JW(I,J+1)+JW(I+1,J) +
          JW(I,J-1))/8
          DO 15 J=1,NJ,NJM1
          DO 15 I=2,NIM1
     15 IUX(I,J)=(2*JW(I,J)+JW(I-1,J)+JW(I+1,J))/4
          DO 20 I=1,NI,NIM1
          DO 20 J=2,NJM1
     20 IUX(I,J)=(2*JW(I,J)+JW(I,J-1)+JW(I,J+1))/4
          IUX(1,1)=JW(1,1)
          IUX(1,NJ)=JW(1,NJ)
          IUX(NI,1)=JW(NI,1)
          IUX(NI,NJ)=JW(NI,NJ)
          END
          SUBROUTINE CONTUR(IZ,NI,NJ,MIN,INT,DATE,LB)
C         ...PRINTS STANDARD NI X NJ GRID WITH CONTOURING
          BETWEEN LINES
C         ...MIN IS THE MINIMUM VALUE, INT IS THE CONTOURING
          INTERVAL
C         ...NJ MUST BE.LE.26
          COMMON/DATA/KALP(16)
          DIMENSION IZ(40,40),LB(5),IPR(34)
          DIMENSION          LINE (126),LIN(26)
          DIMENSION LINE2(126)
          INTEGER DATE
          PRINT 26,LB,DATE
     26 FORMAT (X,6A4///)
          LTOT=INT*16
          NJM=NJ-1
     22 FORMAT (1H)
          NJM=NJ-1
          NIM=NI-1
          NUM=5*NJ-4
          ENCODE (136,900,IPR) (IZ(1,J),J=1,NJ)
    900 FORMAT (3X,26I5,3X)
          CALL PRINT (IPR,34)
          DO 1 IR=2,NI
```

```
      DO 2 JD=1,2
      DO 3 L=1,NJ
    3 LIN(L)=((IZ(IR,L)-IZ(IR-1,L))*JD)/3+IZ(IR-1,L)
      K=1
      DO 4 J=1,NJM
      LINJ=LIN(J)
      LINE(K)=LINJ
      NDZ=LIN(J+1)-LINJ
      DO 5 L=1,4
      K=K+1
    5 LINE(K)=(NDZ*L)/5+LINJ
      K=K+1
    4 CONTINUE
      LINE(K)=LIN(NJM)
      DO 6 L=1,NUM
      JDF=LINE(L)-MIN
    7 IF(JDF)8,9,9
    8 JDF=JDF+LTOT
      GO TO 7
    9 J=JDF/INT
      IS=J-(J/16)*16+1
      LINE(L) = LINE 2(L) = KALP(IS)
      IF (KALP(IS).NE.1H) LINE2(L) = KALP(IS)-37606060B
    6 CONTINUE
      PRINT 902, (LINE(L),L=1,NUM)
  902 FORMAT (1H*,6X,126A1,3X)
      ENCODE (136,901,IPR) (LINE2(L),L=1,NUM)
  901 FORMAT (7X,126A1,3X)
      CALL PRINT (IPR,34)
    2 CONTINUE
      ENCODE(136,900,IPR) (IZ(IR,J),J=1,NJ)
      CALL PRINT (IPR,34)
    1 CONTINUE
      END
```

Appendix II. Fortran program used to calculate atmospheric dispersion of minor constituents based the Pasquill-Gifford method. This program follows that developed by Heimbach (1970) at the University of Oklahoma.

```
C          PROGRAM TO PLOT CODED VALUES OF AIR POLLUTION CONCEN-
C          TRATION, (C) USING THE PASQUILL-GIFFORD MODEL.
C          SY AND SZ ARE THE STANDARD DEVIATIONS IN Y AND Z
           DIRECTION (M)
C          U IS VELOCITY OF WIND WHERE X IS PARALLEL TO V(M/SEC)
C          X,Y, AND Z ARE DISTANCES IN CARTESIAN COORDINATES
           (METERS).
C          Q IS THE RATE WHICH POLLUTION IS GENERATED (G/SEC)
           SOURCE.
C          H IS EFFECTIVE HEIGHT OF POLLUTION SOURCE.
C          NN IS THE NUMBER OF SOURCES, ONLY ONE SOURCE WAS USED
           IN THIS RUN.
C          SSY AND SSZ ARE THE ANGULAR STANDARD DEVIATIONS IN Y
           AND Z DIRECTIONS.
C          REF. P. 125 M.A.E.
C          IS AND JS ARE SOURCE COORDINATES.
           COMMON IS(2), JS(2), X(2,120), Y(2,25), C(1,25,120)
           NN = 1
C          STARTING WITH A GRID POINT AND SOURCE MAKE AN X AND
           Y DISTANCE FROM SOURCE
      1 DO 7 N=1,NN
           READ 2, ISTAB, IS(N), JS(N),SIZE,U,Z,Q,H
      2 FORMAT (3I5,5F8)
           IF (ISTAB.EQ.0) GO TO 6
C          TEMPORARY CHECK.
           PRINT 20, ISTAB,IS(N),JS(N),SIZE,U,Z,Q,H
     20 FORMAT (1H1/3(5X,I5/),5(5X,F10.4/))
           COUNT = 0.0
           JJS=120-JS(N)
C          I=ROW, J=COLUMN,... Y=ROW, X=COLUMN.
C          SAMPLE RUN IS LIMITED TO A FIVE ROW GRID.
           DO 7 I= 1,25
           DO 7 J = 1,JJS
C          -(I-IS)=IS-I
           II=IS(N)-I
C          THE GRID POINTS ARE 100 METERS APART DOWN WIND AND
C          167 METERS CROSS WIND.  THIS ARRANGEMENT IS THE RATIO
C          OF LETTER SPACING TO LINE SPACING ON THE PRINTER.
           Y(N,I) = SIZE*(5./3.)*II
           X(N,J) = SIZE*J
C          CONVERTING SS TO S.
           CALL SCALC(ISTAB,X(N,J),SY,SZ)
           COUNT = COUNT + 1.0
      7 C(N,I,J)=CCALC(SY,SZ,U,Y(N,I),Z,Q,H)
C          A REMINDER, REMEMBER THE DECIMAL POINT ON THE DATA
C          CARD TAKES PREFERENCE.  AT THIS POINT THERE ARE N
C          5X120 FIELDS OF POLLUTION STORED ALONG WITH X AND Y
C          COORDINATES.
```

```
C        TEMPORARY CHECK.
         WRITE (3,22)COUNT
     22 FORMAT (34H THE NUMBER OF ITERATIONS DONE IS ,F8,/73
        HSOME VALUES 0 *F POLLUTION AND THEIR CORRESPONDING
        DOWN WIND DISTANCE ARE )
         N=1
         I= IS(1)
         WRITE (3,21)(C(N,I,J),X(N,J),J=1,15)
     21 FORMAT (1X,E13,6,5X,F8.2)
         WRITE(3,23)SY, SZ
     23 FORMAT(33H THE LAST VALUES OF SY AND SZ ARE ,2(E13,
        5,5X))
         WRITE(3,24) Y(1,3)
     24 FORMAT ( 9H Y(1,3)= ,E13.5)
C        END OF TEMPORARY CHECK.
         CALL CPLOT (NN,SIZE)
         GO TO 1
6        WRITE(3,13)
     13 FORMAT (11H2END OF JOB)
         END

         FUNCTION CCALC(SY, SZ, U, Y, Z, Q, H)
C        PASQUILL-GIFFORD SUBROUTINE
         PIE=3.14159
         A=2.0*PIE*SY*SZ*U
         P1=Q/A
         P2=EXP(-0.5*(((Z-H)/SZ)**2))
         P3=EXP(-0.5*(((Z+H)/SZ)**2))
         P4=EXP(-0.5*((Y/SY)**2))
         CCALC=P1*P4*(P2+P3)
         RETURN
         END
         SUBROUTINE CPLOT(NN,SIZE)
C        POLLUTION PLOTTING SUBROUTINE
         DIMENSION CI(25,120),IC(25,120)
         COMMON IS(2), JS(2), X(2,120), Y(2,25), C(1,25,120)
         CHARACTER IC
C        INITIALIZE ALL GRID POINTS=0.0
         DO 203 K = 1,25
         DO 203 L=1,120
203      CI(K,L)=0.0
         WRITE (3,236) NN
C        CONVERT X≠S AND Y≠S BACK TO GRID COORDINATES, L,K
C        FOR C(N,I,J).  EACH POINT ON THE GRID IS IN UNITS OF
C        100 METERS DOWN WIND AND 166.66 METERS CROSS WIND.
C        NN = NUMBER OF SOURCES.
         DO 201 N=1,NN
         JJS=120-JS(N)
         DO 201 I = 1,25
C        REMEMBER THAT +X IS DOWN WIND.
C        THE 0.5 IS TO TAKE CARE OF ANY TRUNCATION ERRORS.
C        KY = GRID UNITS OFF DOWN WIND PATH.
         IF(Y(N,I))204,205,205
   205 KY = Y(N,I)/(SIZE*(5./3.)) + .5
```

```
          GO TO 202
   204 KY = Y(N,I)/(SIZE*(5./3.)) - .5
   202    K=IS(N)-KY
          DO 201 J=1,JJS
          L = X(N,J) /SIZE + JS(N) + .5
C         ADD POLLUTION FROM SOURCE N TO POINT K,L.
   201    CI(K,L)=CI(K,L)+C(N,I,J)
C         NOW TO PLOT 0 TO * DEPENDING UPON MAGNITUDES OF
          POLLUTIONS.
C         AT THIS POINT THERE IS AN ARRAY OF POLLUTION CONCEN-
C         TRATIONS FOR ONE SOURCE LOCATED AT IS,JS.
C         NOW INCODE CONTENTS OF EACH GRID LOCATION FOR A-
C         FORMAT USAGE.
C         * = E-02 TO E-03
C         9 = E-03 TO E-04
C         8 = E-04 TO E-05
C         7 = E-05 TO E-06
C         6 = E-06 TO E-07
C         5 = E-07 TO E-08
C         4 = E-08 TO E-09
C         3 = E-09 TO E-10
C         2 = E-10 TO E-11
C         1 = E-11 TO E-13
C         - = A CONCENTRATION LESS THAN E-13
          DO 232 K = 1,25
          DO 232 L=1,120
   212    IF(CI(K,L)-1.0E-13)210,213,213
   210 IC(K,L) = 1R-
          GO TO 232
   213    IF(CI(K,L)-1.0E-11)214,215,215
   214 IC(K,L)=1
          GO TO 232
   215    IF(CI(K,L)-1.0E-10)216,217,217
   216 IC(K,L)=2
          GO TO 232
   217    IF(CI(K,L)-1.0E-09)218,219,219
   218 IC(K,L)=3
          GO TO 232
   219    IF(CI(K,L)-1.0E-08)220,221,221
   220 IC(K,L)=4
          GO TO 232
   221    IF(CI(K,L)-1,0E-07)222,223,223
   222 IC(K,L)=5
          GO TO 232
   223    IF(CI(K,L)-1.0E-06)224,225,225
   224 IC(K,L)=6
          GO TO 232
   225    IF(CI(K,L)-1.0E-05)226,227,227
   226 IC(K,L)=7
          GO TO 232
   227    IF(CI(K,L)-1.0E-04)228,229,229
   228 IC(K,L)=8
          GO TO 232
   229    IF(CI(K,L)-1.0E-03)230,231,231
```

```
    230 IC(K,L)=9
        GO TO 232
    231 IC(K,L)=1R*
232     CONTINUE
C       PLOT THE SOURCES.
        DO 206 N=1,NN
        II=IS(N)
        JJ=JS(N)
    206 IC(II,JJ)=1RS
C       WRITE PLOT
        DO 233 K = 1,25
233     WRITE (3,234)(IC(K,L),L=1,120)
        RETURN
    234 FORMAT (1X, 120R1)
    236 FORMAT (1H1,23HTHE POLLUTION PLOT FOR ,I1,27H SOURCES
        INDICATED BY
      * S IS //)
        END
        SUBROUTINE DIST(I,J,WIND,IS,JS,X,Y)
        XX=96.3*(J-JS)
        YY=-120.6*(I-IS)
        STEMP=XX*XX+YY*YY
        IF(STEMP)1,1,2
2       HYP=SQRT(STEMP)
        IF(XX.EQ.0.0.AND.YY.GT.0.0) TEMP=1.570796
        IF(XX.EQ.0.0.AND.YY.LT.0.0) TEMP=-1.570796
C       IF(XX*YY.NE.0.0) TEMP=ATAN2(YY,XX)
        BETA=TEMP-3.1415+WIND-0.437
        X=HYP*COS(BETA)
C       Y≠S SIGN DOESN≠T MATTER IN THE MODEL.
        Y=ABS(HYP*SIN(BETA))
        RETURN
1       X=0.0
        Y=0.0
        RETURN
        END
        SUBROUTINE SCALC(ISTAB, X, SY, SZ)
C       SUBROUTINE TO CALCULATE SIGMA≠S
C       THE CODES 1,2,3,4,5 and 6 REFER TO THE STABILITY
C       CLASSES A, B, C, D, E, F RESPECTIVELY.
        GO TO (1,2,3,4,5,6), ISTAB
1       SY=0.454*(X**0.892)
        IF(X.LE.500.0) SZ=0.191*(X**0.933)
        IF(X.GT.500.0) SZ=0.00025*(X**2.086)
        GO TO 7
2       SY=0.292*(X**0.909)
        IF(X.LE.500.0) SZ=0.124*(X**0.968)
        IF(X.GT.500.0) SZ=0.0423* (X**1.126)
        GO TO 7
3       SY=0.182*(X**0.921)
        SZ=0.109*(X**0.915)
        GO TO 7
4       SY=0.112*(X**0.931)
        IF(X.LE.500.0) SZ=0.113*(X**0.317)
```

```
        IF(X.GT.500.0)  SZ=0.400*(X**0.623)
        GO TO 7
5       SY=0.090*(X**0.915)
        IF(X.LE.500.0)  SZ=0.105*(X**0.773)
        IF(X.GT.500.0)  SZ=0.431*(X**0.566)
        GO TO 7
6       SY=0.063*(X**0.910)
        IF(X.LE.500.0)  SZ=0.113*(X**0.654)
        IF(X.GT.500.0)  SZ=0.182*(X**0.618)
7       RETURN
        END
```

AUTHOR INDEX

SUBJECT INDEX

Air Pollution, 3, 5, 15, 16, 75-97
American Meteorological Society, 3
Arcas Rocket, 12
Brunt-Väisälä Frequency, 38, 66
Chapman Layer, 41
Circulation:
  Tropospheric, 17, 28-40
  Stratospheric, 17, 40-51
  Thermospheric, 17, 51-52
  Boundary Layer, 18-28
Conductivity (Electrical):
  Specific, 58
  Pederson, 58, 59
  Hall, 58, 60
Convection, 35, 36
Data:
  Temperature, 19, 32, 34, 44, 45, 46, 48, 49
  Sunshine, 20
  Relative Humidity, 21
  Precipitation, 23
  Wind, 27, 29, 30, 47, 52
  Ozone, 42
Data Accuracies, 11
Diffusion, 53-56
Diffusion Computer Program, 113-117
Dust Storms, 7
Dynamo Currents, 59, 60, 61
Eddy Transport, 33, 53, 54, 55, 56
Electrical Structure, 57-61
Equation of Motion, 31
Geostrophic Wind Relation, 31
Heat Transport, 33, 54, 55, 56
Hydrostatic Equation, 53
Interpolation Program, 106-112
Inversions, 33, 34, 35
Ionosonde, 9, 15
Jodrell Bank, 52
Loki Rocket, 12
Magnetic Field, 60
Magnetometer, 9
Meteorological Rocket Network, 1, 2, 3, 12, 51
Meteorological Working Group, 11
Meteor Trail Radar, 14, 15, 51, 52
Mixing Depths, 37
Monsoon, 40
National Academy of Sciences, 3
Observations, 8-16
  Surface, 9-11, 18-28
  Balloon, 11-12, 17, 28-40

About The Author

Willis Lee Webb is a research meteorologist with the Atmospheric Sciences Laboratory at White Sands Missile Range and a Lecturer in the Physics Department at the University of Texas at El Paso. He has twenty years of research experience, the first three in the field of cloud physics with the Physical Research Division, Weather Bureau, Washington, D. C. Previously, he worked as a meteorologist at airport weather stations, mostly in the Southwest, with the Weather Bureau and the Air Force. At the Atmospheric Sciences Laboratory he has conducted research in atmospheric acoustics, synoptic rocket exploration of upper atmospheric dynamics, and the earth's electrical structure.

He has published more than thirty scientific articles, and is the author of Structure of the Stratosphere and Mesosphere (Academic Press, Inc. , 1966). He edited Stratospheric Circulation (Academic Press, Inc. , 1969) and Thermospheric Circulation (MIT Press, 1971), both of which are proceedings of scientific meetings which Webb conducted. The first was sponsored by the U. S. National Academy of Sciences through the Committee on Space Research of the International Council of Scientific Unions at Imperial College of Science and Technology in London in 1967 and at the Japan Meteorological Agency in Tokyo in 1968; and the second was a 1970 summer institute in the Physics Department of the University of Texas at El Paso.

Born in Nevada, Texas in 1923, he completed high school there. He attended Arlington State University in Arlington, Reed College in Portland, Harvard University in Cambridge, Southern Methodist University in Dallas, the University of Oklahoma in Norman, and Colorado State University in Fort Collins. He has a Bachelor of Science degree in mathematics from Southern Methodist University and a Master of Science degree from the University of Oklahoma. He is currently working on a comprehensive volume on the global electrification problem, titled Earth's Electrical Structure.